I0518359

Michigan in the War with Mexico 1846-1848: The Cavalry

Paul H. Scheidler

Cover: "An incident of the Battle of Churubusco, August 20, 1847. Sergeant Kenady unloading a powder wagon." 1887 artist unknown. Wikipedia Commons.

To Rollin and Mary Scheidler
The two people who sparked my love of history.

ACKNOWLEDGEMENTS:

This historical monograph would not have been possible without the monumental contribution of LeRoy Barnett and Roger Rosentreter and their work "Michigan's Early Military Forces," along with Kenneth Aubrey Smith in 1950, and Richard W. Welch in 1967, who also attempted to create rosters compiling accurate lists of Michigan men who served in the war. My thanks also go to those who gave assistance over the last decade at the Bentley Historical Library, the Burton Collection at the Detroit Public Library, the Tecumseh Historical Museum and the Historical Library of Michigan. I would like to acknowledge Elden Davis of Howell, a researcher and archivist extraordinaire, for his assistance with obituaries and newspaper accounts regarding these veterans. He, along with Richard Hutchins, Rick Danes, Mark Hoffman, Bob Eliott and Chris Czopek, as well as other members of the Michigan Sons of Union Veterans of the Civil War, helped to make this story possible through their unceasing dedication to the memory of our veterans. My thanks to Cindy Ziegelman, Peggy McCormick-Platz and Mark Lawrence for editing advice. I would also like to thank my wife Amy, who not only lent her ear and patiently listened to my never-ending lists of self-inflicted stress, but lent her eye to my first drafts, for which I am eternally grateful.

TERMS:

When discussing cavalry, or the not-so-familiar title of dragoons, the term "company," "squadron" or "troop" are all used interchangeably to describe the same basic unit of between sixty and one hundred mounted men that were recruited, operated, and fought together. Two of the three terms are specific to cavalry; "squadron" and "troop" are not used when describing infantry or artillery units, just those mounted on horses. Ten of these companies together constituted a regiment, and they are lettered "A" through "K." The letter "J" was omitted for looking too much like "I." Company K is sometimes referred to as "McReynold's Troop" or "squadron," and are sometimes referred to as the "Grey Horse Squadron," although it is more a reference to Captain Phil Kearny's Company. During the period of time when they were serving with Captain Kearny's company of the 1st Dragoons as General Scott's bodyguard, the two groups, getting smaller every week due to losses, soon appeared to be the size of one unit. For that, and other reasons to be discovered, they were often seen and referred to by the rest of the army, and war correspondents, simply as the "Headquarters squadron," or "Scott's Bodyguard Troop."

I have attempted to interpret the pronunciation of Mexican place names as best I can, Americans heard the names and then spelled them phonetically, and regularly wrong. In the spirit of uniformity, I will name them as they named them and as they appeared on American maps and in American newspapers back then.

CONTENTS

LIST OF ILLUSTRATIONS

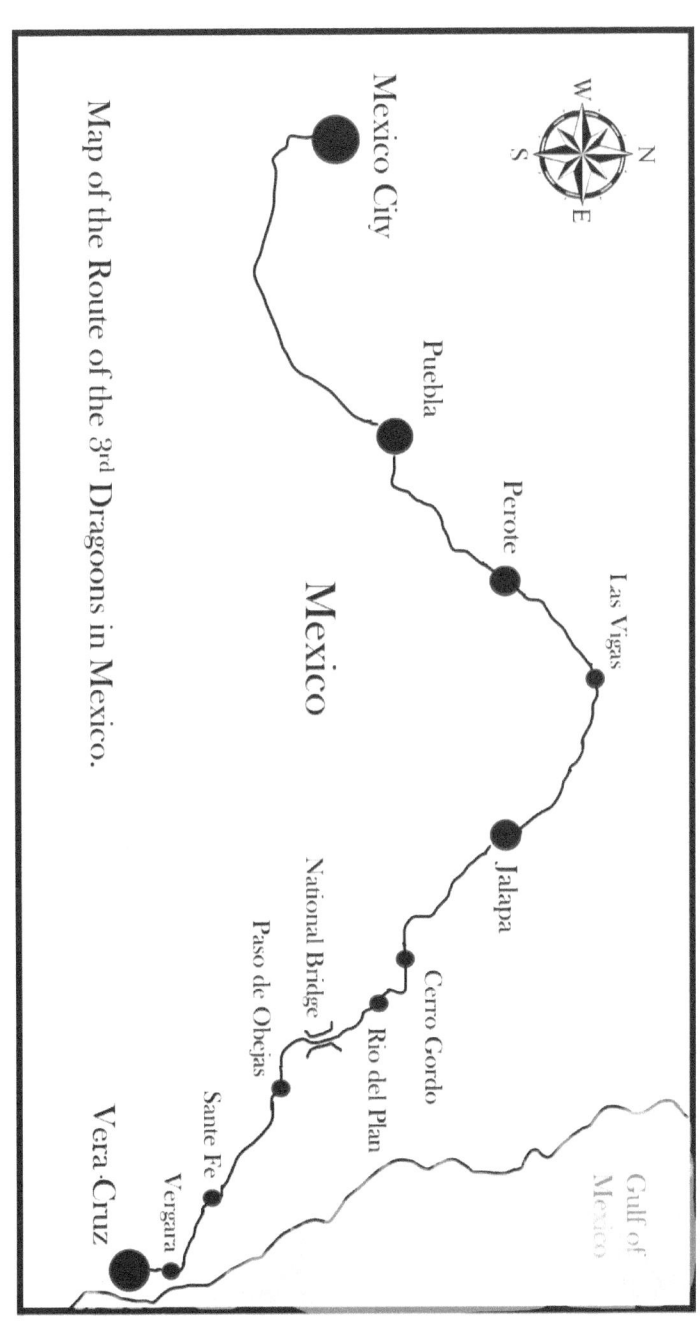

Map of the Route of the **Route** of the 3rd **Dragoons** in Mexico.

Mexico City

Puebla

Perote

Las Vigas

Mexico

Jalapa

National Bridge

Paso de Obejas

Cerro Gordo

Rio del Plan

Sante Fe

Vergara

Vera Cruz

Gulf of Mexico

N
W E
S

1

PROLOGUE

In 1878 a small gathering of aged veterans met for perhaps the last time in Detroit. Many of them walked with a cane, and wore badges reflecting their service, not so different from nowadays. They met in the boardroom of the Detroit YMCA building for their annual reunion.

This was not uncommon in those days, there were tens of thousands of Civil War Veterans in each state, and they met regularly. Newspapers of the time are filled with advertisements for regimental and state reunions. Some of these men were Civil War Veterans, but that was not why they had gathered. They were representatives of a smaller, more unique group of soldiers, from a war that has been nearly forgotten today, as it had been even back then.

At the head of the long wooden table, an aged bespectacled commander sat among them, listening to the members of the welcoming committee and their long, albeit appreciated comments. Neatly dressed and well groomed, he had the dignified air of a professional about him, this made sense since he had spent most of his years practicing law and as a politician. His hair was longer than was common in those days, perhaps to compensate for his receding hairline. He looked down over his spectacles at his coat, unbuttoned it and rummaged through the pockets of his vest. His pocket watch was extricated by a firm tug of the fob, and after a glance returned to its place next to his bottom rib. He bowed his head as the chaplain offered a prayer and took a document from the inside pocket of his jacket. His mind drifted as he pictured the faces of those he had not thought of for many days, and he pushed his chair back as the chaplain finished his prayer and the actual meeting began. He stood, as he had done for

2

many years, and unfolded the paper, looked at it, and then folded it up again as he spoke. He didn't need it; he knew this speech by heart.

Mexican War Veterans at a Reunion (Detroit Evening News)

Thirty-one years ago, on April 26, I marched through this noble Thoroughfare-Jefferson Avenue- to Woodward Avenue, and thence to the river, at the head of 104 as brave and stalwart men as ever marched to the battlefield, and we took passage on the boat on our way to Mexico.

Of the 104 brave men of Company K, Third United States Dragoons, of which company I had the honor to be the captain, not seventeen returned to tell the story, and but five are here with us today.[1]

The paper he was holding shook slightly, but he did not notice. He did not remember how long it lasted, perhaps only for a moment, the people at the table disappeared, and the hot smell of tropic vegetation, dust, and the sulphury smell of expended gunpowder filled his brain. He could see the horse that he was riding, seeming to push through a sea of fleeing humans. They seemed to part before him, disappearing in the smoke, dressed in blue jackets with red trim, white jackets with

---

[1] Andrew T. McReynolds, *Presidential Address to the "Veterans of the War with Mexico.* Pioneer Histories, Volume VI, 1884. 20.

3

red trim, and his sword was coming down on their heads, slashing left and right, over and over again as the fleeing Mexican infantry flung their arms up and spun to the ground, sometimes with a look of terror, sometimes with a look of surprise, and there was the one face, which he could never get out of his head, which gave him a look of cold determination as his sword swung just to the side of it and into the man's collarbone and white leather shoulder strap from his knapsack and bedroll. It wasn't fear, it was a look of defiance, and sheer hatred, and he remembered that he was taken aback at how it affected him. It was that face that haunted him in his dreams. Then he remembered the black open mouths of the cannon barrels, aiming directly at him, and he gasped.

The gasp, and the applause woke him from the glare of the apparition in his mind. He sat, and as the others rose and spoke, he could hear them, but it wasn't clear what they were saying. Instead, his mind saw before it the faces of his own dead soldiers, on the battlefield, in the hospital, and the first one, Private Combs, one of the Tecumseh men, lying on the side of the road northwest of Vera Cruz, coloring the sand dark red with his blood. "So many dead," he thought, "so that we could say something glorious about them in our old age. Was it worth the cost?" The old man removed his spectacles and wiped the perspiration that was forming on his face with his handkerchief. He wanted everyone to believe it was perspiration, far more dignified than tears. Someone was talking about the treasurer's, report and he absentmindedly said, "Aye" as they asked for all in favor to accept it and banged the gavel. A few minutes later he asked if someone else might wish to be the leader of the organization, as he did every year, and stated that he would very much like to no longer be in charge, which was politely refused by the members present. And so, it happened again as it did every year when they got together. Much later that night, as he slept in the hotel room in Detroit, the face of the Mexican soldier came back, as it always did, and glared at him, as it always did.

Andrew T. McReynolds later in life.
(Detroit Free Press, Nov. 28, 1898.)

The old man was Andrew T. McReynolds, and he was the commanding officer of Company K. According to the notes in his pocket, his men were engaged in nine separate "battles" in Mexico. Each company in the 3rd Dragoons Regiment had a slightly different history, since the style back then was to separate the companies for individual service in scouting, convoy operations or with the army for the general engagements. The number seventeen he refers to for the survivors seems hyperbolic. It appears to have come from the days

5

after the badly wounded McReynolds left Mexico with his arm in a sling and was on his way back to Michigan to convalesce. In an interview with a reporter on his way home, he related that when he left Mexico City, "there were but seventeen fit for duty in the company.[2]"

Seventeen of one hundred and four? What disaster befell them? What really happened to these young men from Michigan, according to the evidence available? They did not simply disappear, they can be tracked, to an extent, and their story can be brought into the light, and some explanation gained. In addition, there were more than the original one hundred and four. According to the enlistment and pension records there were one hundred and thirty-nine individuals who could attach the designator "Company K" to their military service.[3] When Captain McReynolds left Michigan, he marched off with one hundred and four, so where did the others come from? Another mystery to be solved.

What caused this misadventure? The trouble started after Texas gained its independence in 1836. Mexico did not accept it and claimed that one day their honor would be restored, and they would get it back. In the meantime, if the United States ever annexed Texas, that would mean war. And then it did. And so, Mexico did what they said they would do. To add insult to injury, President James K. Polk asked the Mexican government to sell the northern provinces of New Mexico and California to the United States, and sent an "Army of Occupation" under General Zachary Taylor to Texas to make sure the Texans were protected. This was just another slap in the face to the Mexican government, and a strong spirit of nationalism and revenge swept the capitol city. A large Mexican army was sent north to the disputed border between the two nations, and by late April and early May 1846, the two armies met, and blood had been spilled, and the two countries were at war.

In the early days of 1846, when hostilities began, the United States Army only had two cavalry regiments, the First and Second Dragoon Regiments of approximately 600 men each. They seldom operated as

---

[2] *Hillsdale Whig Standard,* December 21, 1847.

[3] Barnett, LeRoy & Roger Rosentreter, *Michigan's Early Military Forces,* Wayne State University Press, Detroit, 2003. Chapter Six.

6

complete regiments and manned the lonely outposts of the west in isolated detachments of companies of about 60 men each, depending on location and importance of whatever area they were ordered to patrol. As dragoons, they were able to function as heavy cavalry with dragoon saber and pistol, if necessary, but they could also function as infantry when dismounted, by using the short carbine rifle or musketoon they had slung across their backs or attached to their equipment. In this way they could hold a critical position with the mobility to escape quickly if required but also had the heavy hitting power of a charge with sabers drawn, especially in the pursuit of a routed or broken enemy. They were perfectly suited for escorting wagon trains headed for Oregon or traveling vast distances to keep the peace with the Native Americans along the American frontier, there just weren't very many of them.

As political bargaining fell apart and the military of both sides began to maneuver against the other along the southern border, the shortage of trained horsemen became clear. For the most part, most companies of the First Dragoon Regiment were occupied with the movement west, strategically occupying Santa Fe, and then heading for California once war was declared. This left only one cavalry regiment, the Second Dragoons, to support the American Army in Texas under General Taylor. Scores of independent volunteer companies of "cavalry," "mounted infantry," and "Texas Rangers," appeared from various states during the first months of the war, but they had varying degrees of usefulness. The underlying shortage became clear in the first battles along the Rio Grande in May of 1846. The Americans had two regiments of cavalry on hand when the war began, the Mexicans had seventeen.

CHAPTER ONE

From Detroit to Vera Cruz

The men are pleased with all of our officers. They are fine fellows, and if they and the boys under them do not give a good account of 'The Michigan Dragoons,' it will be because we have no chance to show what we can do.
-Allen T. Welch, The Detroit Democratic Free Press,
May 8, 1847[4]

Although the first two battles of the war at *Palo Alto*, and *Resaca de la Palma* in 1846 both resulted in victories for the Americans, they also laid bare the weakness of the American Army compared to the Mexicans, besides the fact that it was very small in comparison. The most glaringly obvious problem was that the Americans suffered from a severe lack of cavalry. The emergency contingent "Ten Regiment Bill" which Congress passed in February of 1847 planned to fix the shortage with a third regiment of dragoons, along with nine other infantry regiments, an enlargement of the artillery regiments, and additional officer positions assigned to existing units. Although they would be enlisting for the war, they would be considered "regulars" not like the volunteers, who were considered short term resources by the regular army and found to be generally more trouble than they were worth. Many one-year militia units had gathered and headed off to Mexico from across the country in 1846 when the war was new. These volunteers had fared badly, lacking discipline and leadership,

---

[4] Allen T. Welch, *Detroit Democratic Free Press*, (Cincinnati, OH.) May 8, 1847.

8

and they paid the consequences with the lives of their men.

The legislation affected Michigan in several ways. "In the Spring of 1847, Michigan was called upon to raise three companies of infantry and one company of dragoons. Since these troops were to be part of the regular army, the 15th U. S. Infantry Regiment and the 3rd U. S. Dragoons, the officers were appointed by, and received their commissions from, politicians in Washington D. C."[5] The appointment of officers came first, and all over the country young men vied for letters of appointment through governors, senators, and congressmen of their states for these "Regular Army" positions. The young men appointed to officer positions within these regiments had reasons for doing so. There had been no major military conflict since the War of 1812, and the heroes of that war had gone on to become the prominent men in their communities, even President of the United States. They saw this war as a mechanism to move up the social and political ladder in their communities. Different states furnished individual companies to the dragoon regiment, and they came from all over the United States, primarily east of the Mississippi. The State of Michigan was tasked with filling Company K of the 3rd Dragoons. The officers were chosen from the elite families in Michigan at the time, especially the sons of famous military and political leaders from the last war. One of those that received a commission from the government became an officer on the regimental staff, and the remainder were commissioned as company grade officers.

The most senior Michigan man belonging to the regiment was Lewis Cass Jr., the thirty-one-year-old son of the famous Senator Lewis Cass, a former soldier and politician from Detroit. Cass was appointed as one of the two majors authorized as part of the "Field & Staff" headquarters of the regiment. He served on Colonel Edward G. W. Butler's staff and was second in command of the 1st Battalion of the 3rd Dragoons.

Captain Andrew Thomas McReynolds was an Irishman, who came to Detroit as a young man, and became a lawyer and until recently, a Democrat in the State Legislature of Michigan. He was

---

5 Kenneth A. Smith, "Michigan's Military Participation in the Mexican War." M. A. Thesis, Wayne State University, 1951. 19.

assigned as the commander of Company K. He had some experience in organizing and leading troops in the famous "Toledo War," and had a hand in raising militia in the Detroit area during the Canadian crisis known as the "Patriot War," and now for the first time he would lead troops for the federal government, it would not be the last.

Under McReynolds there were three Lieutenants serving in the company, one first lieutenant and two second lieutenants. First Lieutenant John T. Brown Jr. was a twenty-eight-year-old officer from Tecumseh. His father was General Joseph White Brown, and the name and reputation of his father, a general from the War of 1812, helped him gain the senior position, along with the fact that Lieutenant Brown was responsible for bringing thirty-three recruits he had already gathered from the Lenawee County area to the rendezvous at Detroit. He had attempted to create the first "Flying Battery" of Horse Artillery[6] in the state, but with little success, since no one in the state had asked for such a thing. Brown petitioned for them to be included in Company K, and they were quickly accepted.

Second Lieutenant John C. D. Williams of Detroit. Born September 2, 1823, he served with distinction in the unit. His father was also an officer in the War of 1812, and was one of the early mayors of Detroit, the Honorable John R. Williams.

One officer from Wisconsin Territory was also assigned to the company. Second Lieutenant Francis Henry was at Mineral Point, when he received word of his commission and accepted it as a Second Lieutenant in the 3rd Dragoons. He was originally from Galena, Illinois, and he reported his age as twenty when he wrote back to the War Department on March 26, 1847. The newly minted officer added a "P. S." at the bottom of his acceptance letter. "I cannot report myself to Captain Andrew T. McReynolds as I am at present uninformed of his place of residence."[7]

These freshly commissioned officers were given enlistment forms

---

[6] "Flying-Artillery" meaning a battery of four or six cannon with two-wheeled artillery limbers pulled by four or six horses each was a relatively new concept in those days. The ability of American field artillery to "fly" rapidly across the battlefield and deploy massive firepower in key locations was one of the deciding factors in the war.

[7] NARA, Francis Henry Letter to the War Department, March 26, 1847. (H180.)

and a small amount of money to travel with and they set off across the counties of Michigan. Recruitment for the company began across the southern half of the Lower Peninsula in February and ended in Detroit during the last week of April 1847. Each officer was sent out to a different part of the state (except Henry) to put up recruiting posters and set up temporary recruiting offices in small towns and villages.

Why would young men from Michigan enlist in these units, and agree to give up their freedom for the "duration of the war?" Why sign up to travel to a foreign land and take part in a struggle that was not geographically significant to them? The answer was primarily the land warrant for 160 acres of farmland issued by the government to any volunteer that successfully completed their enlistment, and as a compensation for the spouse of any soldier or their family if they did not return. There was also a degree of patriotism and a sense of adventure that seems to exist at the beginning of most wars by a generation that has not lived through one. Many of the men who enlisted were immigrants and had not yet established themselves in the new state. The army pay and land would allow them to start a new life in Michigan. There were some special demographics within this group of men who raised their right hand and volunteered. One third of the men were from Tecumseh, and probably all knew one another. The other two-thirds were from Detroit or across the river in Windsor, Canada. Also, about one third were recent immigrants, and not native-born to the United States. They tended to be Irish or German.

Another notable demographic was how many of them were brothers, some were even sets of twins. Most recognizable among the Tecumseh men were the three Ellis brothers, Avery, Commodore and William, ranging in age from twenty-one to twenty-five. There were several sets of brothers from the Detroit area. Enos and Joel Parish; Isaac and William Gibson; and one set of twins, Alden and Gilbert Ball; both aged twenty-one, also from the Detroit area.

Much was made of the fact that the officers were not to accept anyone less than six feet tall, but the enlistment records clearly reflect that many exceptions were made.[8] Although recruiting was going well, Captain McReynolds was suffering logistics issues of his own.

---

[8] Silas Farmer, History of Detroit, Wayne County, and Early Michigan, 3rd ed. Revised, (Detroit, 1890), I, 303.

11

The rendezvous location was not Fort Wayne, which seems logical for Detroiters, but that is because it was still under construction at the time. Instead, they took up residence in the old "Detroit Barracks" buildings at the corner of Catherine Street and Russell. Most of the current military footprint in the Detroit area was located at the Dearborn Arsenal, which was miles away, so the logistical support at the barracks was minimal at best. McReynolds had a lot to do: uniform and equip the soldiers, arm them, feed them, pay them, and above all train them to be cavalrymen. Due to a lack of equipment and supplies available in Detroit, he sent a letter to Colonel Butler at his higher headquarters on March 29, 1847, requesting to move to New Orleans as soon as possible once his numbers were filled. He had purchased bed sacks, sheets, and blankets for his men out of his own pocket, as he expected these to be available at the barracks and they were not forthcoming. He also mentioned that he was as yet unaware of how successful his lieutenants were at recruiting, except for Second Lieutenant Williams, who reported that he had been lying sick in bed in Buffalo, New York, and unable to move.[9] The longer that McReynolds stayed in Michigan, the longer he would need to be locally supplied, which was problematic at best. He was right in surmising that the sooner he got to New Orleans or Mexico itself, the sooner he could draw from the vast resources of the federal government.

The people of Detroit were split on supporting the war, the Democratic party of President Polk was all for it. The opposition, formerly the "Federalists," then known as the Whig party, believed it was simply a war to expand slavery to new territories and therefore tried to vilify it. The competing political parties used any good or bad treatment of soldiers gathering for deployment as an example to further their own agenda. The Detroit newspapers wanted to put their "spin" on the proceedings, and those for and against the incumbent political party had their say.

> Captain [Mc]Reynolds of the 3d Dragoons is recruiting rapidly, having himself accepted over 50

---

9 NARA, Andrew McReynolds, Letters Received by the Adjutant General, 29 March 1847.

first rate men at this post. He will have his marching compliment early next week, and would have had it weeks ago but for the absence of some of his Lieutenants, and the neglect of the proper department in forwarding the 1st Lieutenant the necessary papers.

-The Detroit Advertiser[10]

We understand from Captain McReynolds that his company is full, and that his first Lieutenant has enlisted his full quota of men. The charges of neglect of the War Department in furnishing the necessary papers, made by the federal organ is, therefore, all moonshine.

-The Detroit Democratic Free Press, April 12, 1847[11]

Company K was also singular in that it boasted among its members at least five enlisted men who referred to themselves as "The Printers." Gentlemen from several different newspapers that volunteered to send back the exploits of the unit. They could be described as an early form of being "embedded." The Detroit Democratic Free Press reported that,

Among the recruits are several printers, of course. We say, of course, for who ever heard of fighting to be done, without printers having a hand in it, and as a class better and braver soldiers never shouldered a musket. The "Free Press" will be represented in the Dragoons by Allen T. Welch, and Daniel Cruice, -the "Advertiser" by Thomas Burnham, and the Ann Arbor "True Democrat" by William Patten. When the 'revels' so long talked about in the "Halls of Montezuma" really commence, the "Printer Dragoons" will be on hand there, and at every other place within reach, where fun or fighting is to be had.[12]

---

[10] Detroit Advertiser. April 12, 1847.

[11] Detroit Democratic Free Press. April 12, 1847

[12] *Detroit Democratic Free Press*, April 24, 1847.

13

The newly minted cavalrymen were mustered into federal service at the Detroit Barracks on April 22, 1847. They were most likely equipped with the Model 1843 Dragoon Saber, and the Hall Model carbine if they were available from the Dearborn Arsenal, if not they were issued a flintlock or percussion cavalry musketoon. Few men saw a revolver before the war ended, they were so new, and most were issued a percussion or flintlock single shot pistol.[13] They may have been issued some or all of the basic uniform of a cavalryman before they left Detroit, but this cannot be confirmed. The officers would have all been uniformed correctly as they had to pay for their own uniforms. The amount of training they did is unknown. Whether they brought their own mounts to Detroit or not is also unknown, but likely. What is known is that they did not leave with them, when the time came, their horses were left behind. It was expected that the government would furnish mounts when they arrived at a supply base in New Orleans or when they got to Mexico. The following excerpt from a Detroit newspaper described the company as it prepared to depart:

> The Michigan Company of Dragoons under the command of Captain A.T. McReynolds of this city, will leave this morning for Point Isabel[14], and we shall be disappointed if they be not the brag company of the regiment. The men are composed of the best material for service. Young, proud and high spirited -many of them with talents, character and education which would fit them for high stations- impelled by ardent patriotism, they leave their homes to fight the battles of their country, and nobly will they uphold the honor of the Republic.[15]

---

[13] Ron Field, "Mexican American War 1846-1848." 47-48; Randy Steffen, The Horse Soldier 1776-1943, 127.

[14] This was inferred, since that was where the headquarters was going, but not accurate. Orders were already being made to split the unit and send half of it to General Scott in Vera Cruz.

[15] "Presentation of a Sword to Captain McReynolds." *Detroit Democratic Free Press*, April 30, 1847. They had stopped in front of old General Brady's house. He came out and reviewed them.

According to the Detroit Advertiser and Captain McReynolds, the company (dismounted) marched down to the docks and left the city of Detroit on Monday morning, April 26th, 1847. They were to take a steamship to Toledo, a canal boat down to Cincinnati, and from there a river steamboat down to New Orleans.

Before their departure a small ceremony occurred, presumably at the foot of Woodward Avenue at the docks as Company K was about to board. The Free Press wrote about it afterward.

> As they marched from their quarters to the boat which awaited them, their fine appearance and soldierly bearing elicited the warmest admiration from the vast numbers who attended to bid them an affectionate farewell. Our country's flag will not be defended by a finer or braver set of men. Michigan may well be proud of them and feel assured that they will fully represent her valor and patriotism. They have already won, from one of the oldest and bravest veterans in our service, the flattering remark, 'They are the finest body of men that I have seen since the last war.' As the company embarked, the wharves and boats in the vicinity were crowded with citizens. The committee of arrangements, consisting of Lieutenant Colonel A.S. Williams, Thomas Gallagher, F.C. Alcotte, Sen., E.N. Wilcox, Colonel Flood, Captain O. Callaghan, and William Grey, attended for the purpose of presenting a sword to Captain McReynolds.[16]

The company traveled for six days from Toledo before reaching Cincinnati by way of the Miami and Lake Erie Canal. As promised, the printer, Welch, described the trip in the Detroit Democratic Free Press.

> On our way down, our captain gave us liberty to leave the boat and go through every town on the

---

[16] Ibid. It is worth noting that several of those presenting the sword would themselves be deploying in uniform before the year was out.

15

canal. At Defiance, we visited the old fort, erected by General Wayne during the Indian War, after the defeat of General St. Clair, and to which old "Mad Anthony" gave the characteristic name of Fort Defiance.

He also described how,

> -the boys would rally out in search of 'eatables' and bring in chickens, turkeys, eggs, and almost everything else in the eating line, which they say they paid for. Some I know did purchase articles for a price that would make the Detroiters stare, provisions were so cheap.

The company arrived in the city of Cincinnati the following Saturday, the evening of May 1, 1847, by way of the Miami Canal. A local paper reported that there were 106 of them, and that McReynolds, Brown, Williams, and Henry were all with them.[17] They had left within a day or so, claiming that they were to embark for Point Isabel, Texas. under Major Lewis Cass Jr., but not Company K.[18]

From this we can infer that they believed they would be deployed to the northern theater of operations under General Taylor. Part of the 3rd Dragoon regiment was sent there, the First Battalion,[19] Welch also stated in his letter to the paper that they planned to leave Cincinnati on Monday, May 3, 1847, for New Orleans. Smith reported that they reached New Orleans in early May.[20] Although very little is known about their time in New Orleans, it is known that they were still there

---

[17] The term for them is "Aggregate." Most units in Mexico brought a few civilians along, some servants, some laundresses, or hospital matrons, who were married to a member of the unit, and in the case of southern units, slaves. This accounts for the 106 instead of 104. The author has discovered the names of two such women, however, we do not know the names of most of these anonymous civilians.

[18] *Hillsdale Whig Standard*, May 18, 1847.

[19] If a regiment was to be split in half, the separate halves were referred to as a "battalion."

[20] Kenneth A. Smith, "Michigan's Military." 29

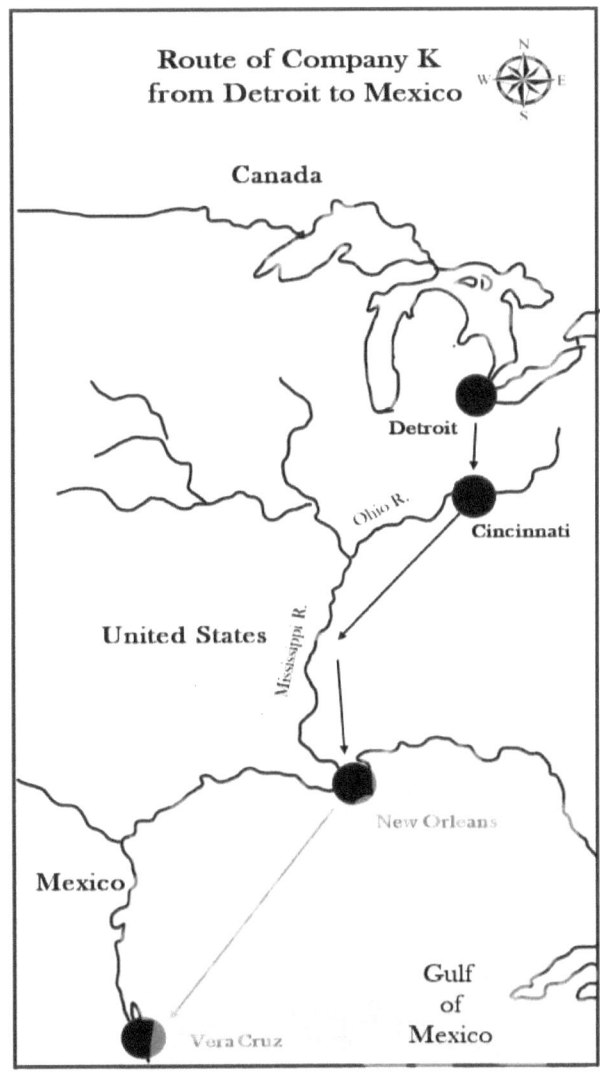

Route of Company K from Detroit to Mexico

Canada

Detroit

Cincinnati

Ohio R.

United States

Mississippi R.

New Orleans

Mexico

Gulf of Mexico

Vera Cruz

on May 15, 1847, because the company reported that Private Daniel Allen, a thirty-year-old carpenter from Tecumseh, deserted there on that day. What story he developed to explain his sudden return to Tecumseh alone is unrecorded.

In their haste to get to "The Seat of War," as Private Welch and other writers of that time often referred to Mexico, the different companies of the different states all traveled alone, got to New Orleans at different times, and left the United States in small compliments of one or two companies together. All had a different story of how they got to Mexico, where they ended up and what they did. There was a reason behind all the chaos.

A significant event occurred while the regiment was forming in New Orleans. The 3rd Dragoons were unceremoniously divided into two battalions, in other words split in half. Five companies were sent to Zachary Taylor's Army in Northern Mexico under Major Cass, with the Field and Staff under Colonel Butler joining them there. The remaining five companies were tasked to sail south under the unit's second-in-command, Lieutenant Colonel Thomas W. Moore. Company K went with the Second Battalion further south to join General Winfield Scott's army at *Vera Cruz*. The companies in the south were generally unaware of what those with Taylor's Army were doing during most of the war, and vice versa.

In fact, the 3rd Dragoons may have never formed as a complete regiment in one place. If it did, it may not have been until it was preparing to discharge its companies at Jefferson Barracks after the war in 1848. The early months of its existence were driven by two conflicting realities. The army wanted the unit in both theaters of war as quickly as possible, and that there was neither enough equipment nor were there enough horses to make them useful. Every general was desperately requesting more cavalry. On the other hand, the companies were often rendezvousing and marching off without most of what they needed to fight. Captain Lemuel Ford, commanding Company D from Indiana, wrote of his difficulties once he got to New Orleans. Those difficulties suffered by McReynolds would have been the same.

I arrived here on Sunday the 16th with 83 men and have received one since my arrival: in all 84.

I have 6 in hospital and 2 sick in camp. I have received 80 sets of woolen clothing, 160 pair drawers, 160 pair flannel shirts. 100 pairs of boots the half of them unserviceable to my men on account of there being too small. 80 forage caps, 80 haversacks, 80 knapsacks, 80 tin canteens with straps, 160 pair stockings, 80 blankets. 14 common 2 by 2 wall tents with flies and poles, 55 great coats. We want more clothing of every description, several of my men lost their caps on board the boat. We have no cotton clothing.

Several of my men are without shoes. We have neither horses nor horse equipage for dragoons.

-Lemuel Ford, Captain 3rd Dragoons."[21]

As the new commander of a unit which he had never seen, nor would be seen all together, Colonel Butler had his hands full. The War Department was making decisions that would split his unit up and he was desperately trying to keep his regiment together. The confusion is apparent in a letter written to his superiors in Washington a few days before he left New Orleans.

General:

Your communication of the 11th Instant reached me yesterday, and I cannot express to you the regret and mortification which the president's determination to separate the two wings of my regiment has caused me, as I look upon its efficiency and opportunities for distinction as thereby reasonably destroyed.

The companies of Captains Ford, Duperu and McReynolds depart this afternoon for Vera Cruz. Captain Gaithers company embarks for the same point, via Point Isabel, in the hope of finding there, its deficient clothing. Orders have been sent to Captain

---

[21] NARA Captain Ford's Letter to Colonel Butler, May 19, 1847.

Sitzgreaves to sail directly to Vera Cruz.

Lieutenant Colonel Moore goes with the companies for Vera Cruz, and the 2nd major of the regiment, when he reports himself will be placed under his orders.

Major Cass is with three companies at Palo Alto where he intends to remove his headquarters by the 1st of June.

Colonel Butler[22]

Captain John S. Sitzgreaves never went to Vera Cruz; he ended up with Colonel Butler in General Taylor's Army. Captain William H. Duff's Company E from New York went instead, and the unfortunate captain died at Vera Cruz shortly thereafter. Colonel Butler wrote a description of the shortfalls as he prepared to leave New Orleans and join General Taylor's half of the army on June 3, 1847, he stated that the regiment, "has neither standard, guidons, music nor farriers."[23]

Lieutenant W. G. Moseley, a Louisianian of Company G., recounted the journey to Vera Cruz many years later for a Veteran's magazine. The first four companies of the regiment that were to join Scott's army under Lieutenant Colonel Moore were designated as the Second Battalion of the 3rd Dragoons, but to anyone serving with General Scott's Army they were simply "The 3rd Dragoons." They consisted of Company G under Captain Alphonse M. Duperu from Louisiana, Company C under Captain Edgar P. Gaithers from Kentucky, Company D from Indiana under Captain Lemuel Ford, and Company K from Michigan under Captain A. T. McReynolds, but even that organization was fleeting. They did not know that a fifth company, Company E belonging to Captain Duff, would be following in a few weeks. Lieutenant Moseley wrote about his company and the others when they boarded the transport steamer *Fashion* and headed south. Moseley stated that, "the voyage was flat, and without incident."[24]

---

22 NARA Colonel Butler's Letter to HQ May 20 New Orleans.

23 NARA Colonel Butler's Letter to HQ June 3 New Orleans.

24 W. G. Moseley, "Reminiscences of a Mexican War Campaign." *The Vedette*, (January-June 1880): 1.

The crossing time from New Orleans to Vera Cruz was approximately four days, based on data collected from the other Michigan units and journal and letter information from the period. As the *Fashion* came to a stop at Vera Cruz, the four companies of the battalion disembarked from the steamship and were transferred by "surfboats"[25] in the harbor to the stone pier between the stone Fortress of San Juan de Ulloa and the looming city walls. These boats were large, flat-bottomed landing craft specially built for the landings that occurred in March of 1847. After receiving directions, the companies marched through the city on foot to the north, out the northern gate and onward to the seaside village of *Vergara*, which was located just north of the city on the beach of the Gulf of Mexico. It was from there that the royal road led north and west toward the National Bridge, before turning west toward the cities of *Perote, Jalapa* and then *Puebla*, which was where Scott's Army lay at that moment. The men of the four companies waited at Vergara for some time as they had no horses for themselves and no transport for their equipment and baggage.[26] The sight and scale of activity, along with the chaos of thousands of troops and supplies in a state of constant movement, must have presented quite a spectacle.

The American Army under General Scott had been building up at Vera Cruz since the landing on March 9, 1847, and by the time Company K arrived almost two months later, a massive logistics base and debarkation point had blossomed along the beach south and north of the city. This, however, did not appear to help the Michigan boys with their equipment shortages. Lieutenant Colonel Moore wrote to Colonel Butler on May 25, 1847.

> The detachment [Second Battalion] under my command arrived here this morning and with the exception of a few cases of measles and seasickness (eleven in all) in good health. The quartermaster says that they have no dragoon clothes or equipment,

---

[25] These were the original landing craft General Scott had created for the landings back in March, and the survivors were now clearly being used for shuttling troops to the docks.

[26] W. G. Moseley, "Reminiscence," 1.

saddles, bridles, halters, etc., etc., and that one hundred of his best horses have been assigned to a volunteer captain, who, with his small detachment, started for the upper country, and rumor has it, has been cut off; that the remaining horses here are all required to remount the 2nd Dragoons and for artillery purposes. We will of course be detained in this vicinity until horses are sent. I need hardly add, that I am sure of your cooperation in relieving us from our very unpleasant condition.[27]

The men pitched their tents if they had them, and lived near the beach for approximately eleven days as they attempted to fill all of their missing equipment lists and at least present the look of cavalry. One member of the regiment wrote.

The encampment on the beach was a practicing school for patience and soldierly fortitude, that would have made a martinet smile. A sudden tempest of wind and rain (always at night) was sure to prostrate tents, stampede horses and create a general "clamjamflamery" that was perfectly maddening and ludicrous. It was there I learned to sleep in falling rain, under water-soaked blankets; in fact, in a genuine hydropathic swathe.[28]

Signs that they were getting ready to leave Vera Cruz soon began to appear. By June 3, 1847, Private David B. Ousterhout, one of the Tecumseh boys, was left in the general hospital there on city waterfront that day, unable to make the trip.[29] Whether he recovered and rejoined the unit or was immediately sent back to the United States is unknown. He was discharged at Jefferson Barracks when the

---

[27] NARA, Lieutenant Colonel Moore's Letter to Colonel Butler, May 25, 1847.

[28] Moseley, 1.

[29] If one wanted to tour the same hospital building today, one only needs to make a reservation at the Veracruz Holiday Inn downtown, it still exists.

regiment mustered out the following year so he must have rejoined the company at some point. That same day, Colonel Butler and the headquarters of the 3rd Dragoons boarded ship and left New Orleans. They headed for Point Isabel in the Northern Theater of Operations, to work for General Zachary Taylor.[30] Even at that point some companies of the regiment had not completed recruiting, nor had they traveled as far as New Orleans.

Back in April of 1847, General Scott's Army was organized into separate marching divisions with long wagon trains and began its winding march from Vera Cruz toward Mexico City. They traveled along the National Road (sometimes called the "Royal Road" originally built by Cortez) that wound its way up from the coastal areas toward the highlands and mountains that lay between the sea and Mexico City. Scott needed to get the army out of the low coastal areas and up into higher cooler climates before the yellow fever season, known locally as "el vomito," destroyed his forces without a shot being fired. Before Company K arrived, General Scott's forces had fought General Santa Anna's Mexican forces at the Battle of Cerro Gordo on April 17th and 18th and then pushed on past Jalapa, Perote and on to the City of Puebla, where it halted in order to reconsolidate and wait for reinforcements.

General Scott did not have enough troops to move on toward his objective, Mexico City, especially since many of his volunteer units, having completed their one-year enlistments, were allowed to march back to the coast and head home. He was desperately short of mounted cavalry to patrol the whole distance from Vera Cruz to Puebla, so he chose to cut himself off from his long supply line and leave only a few defended garrisons along the path back toward Vera Cruz and the coast to consolidate his forces. The Mexican Army took advantage of this by constantly cutting the road and forcing wagon trains to fight or turn back. Company K found itself involved in these fights almost immediately.

---

[30] NARA Colonel Butler's Letter to the Adjutant General, Washington D. C., June 3, 1847.

CHAPTER TWO

First Contact with the Enemy.

*We were ordered to charge upon the enemy who
attacked the rear which was done in good style.
-S. P. Cargill[31]*

On Thursday July 8, 1847, the people of Detroit opened their morning papers to find that the boys of Company K had been in their first major battle, really a series of running fights between Vera Cruz and their destination, the city of Puebla. Such was the difficulty of getting letters out of Mexico, that the battles had happened a month before the people of Detroit knew about them. Weeks earlier, on June 9, 1847, Private Samuel Pidesco Cargill took pen in hand while still fresh from the road having just returned to a location near Vera Cruz. He had been part of a wagon train that had failed to break through Mexican forces that were blocking the National Road. The perilous single supply line from General Scott at Puebla to his supplies at Vera Cruz was a dangerous route and General Santa Anna had sent regular and irregular forces to continuously cut General Scott off from the coast. Their orders were to attempt to trap and destroy any convoy trying to make its way to Puebla. So long as the road was blocked, Santa Anna had the opportunity to destroy Scott's smaller and isolated army stranded halfway to Mexico City. Nevertheless, when General

---

[31] Samuel P. Cargill, *Detroit Democratic Free Press*, (Vera Cruz, MX.) July 8, 1847.

24

Scott could get a message back to Vera Cruz, he ordered the leaders of the arriving troops to form protected convoys and attempt to fight their way through.

Colonel James S. McIntosh, a Regular Army infantry officer who had also seen action in the War of 1812, formed a wagon train protected by three dragoon companies; Captain Ford with Company D and Captain McReynolds with Company K as well as a part of Company G that was walking, as there were not enough horses for all. According to his report, McIntosh also had six companies of infantry, one of them being from his own unit, the Fifth Infantry Regiment, in all he estimated, "six hundred and eighty-eight men, eighty-three of whom were sick. Nearly all these men were recruits, and of course knew little or nothing of their duties."[32] There was most likely a company or two of the Fifteenth Infantry Regiment and also a company or two of the Fourth Infantry Regiment.[33] McIntosh reported that he left Vera Cruz with one hundred and twenty eight wagons, (Cargill estimated about two hundred wagons and four hundred pack mules carrying "a large amount of quartermaster and ordnance stores,") five hundred and twenty-eight draft horses and mules, and four hundred pack mules, and was approximately three-miles-long when the column departed Vera Cruz on June 4, 1847, heading through the sandy part of the road near the coast with great difficulty. Walking alongside the wagons in the center were the infantry companies. Captain Duperu of Company G of the 3rd Dragoons had decided to send half the men ahead without mounts, and so the Louisiana men under First Lieutenant Moseley were also walking along the right side of the wagons, acting as foot soldiers for the time being. Captain Duperu and the other half of Moseley's company were left behind at Vera Cruz, presumably to bring the horses forward once they became available.[34] Opposite them, on the left side of the wagons, Captain Joseph H. Whipple,[35] with forty men of his company from the

---

[32] Hamersly, L. R. Ed., "General Lane, Our Cavalry in Mexico." *The United Service Magazine.* (June 1896): 487.

[33] Ibid, 489. He mentions an officer from each regiment with their men.

[34] NARA Colonel Wilson's to the War Department June 9, 1847.

[35] Captain Whipple died two weeks later on June 30, 1847, of disease in Perote.

Fifth Infantry Regiment, were walking along as well.[36] According to Cargill, it was common knowledge in Vera Cruz that this train was also carrying a large amount of gold, 350,000 dollars in specie according to McIntosh's report.[37] It was to be used to pay the troops and this was the reason for the heavy attacks by Mexican forces. It did not help that several sources claimed the local English-speaking newspaper had published this fact in the previous day's paper at Vera Cruz!

They were already exhausted from the heat and terrain when they came under fire. According to Colonel McIntosh they had only managed three miles on the first day, nine miles the second. One soldier was reported to have simply fallen dead from the excessive heat. The wagon train was wildly uncontrollable since they had attempted to hitch unbroken mustangs to the wagons, and half the mule drivers did not speak English. Moseley stated, as did almost everyone in that war who wrote about that road, that,

> For the first ten miles the road is a bed of deep burning sand, glowing, sweltering, and blistering under the vertical rays of a tropical sun. Lined on each side with an impenetrable thicket of the most profuse vegetation, bristling and repulsive with thorns and briars of mammoth size, and impervious to the breeze.[38]

The first attack was made just after sun-up on the second day out, when they had only marched about twelve miles north of Vera Cruz. The Mexicans opened fire upon the dragoon company under Captain Ford,[39] whose horsemen had been screening and scouting at the front of the train. Private Cargill stated that "The attack was made by a gang of Rancheros, of about 100, who fired from a chaparral.[40]" Lieutenant Moseley, who was trudging alongside the wagons on the right-hand

---

[36] Lane, Our Cavalry in Mexico. 489.

[37] Ibid, 487.

[38] Moseley, 2.

[39] Captain Lemuel Ford of Indiana, age 49, Company D.

[40] Cargill, *Democratic Free Press*, July 8, 1847.

26

side of the convoy, remembered it this way:

> Suddenly, like a clap of thunder on a cloudless day, a withering volley is poured into our flank. Taken by surprise, utterly confounded, and stupefied by the ringing shouts and whistling balls, I stood rooted to the spot for a few moments. When, on looking about me, I found myself almost alone in the road, a prominent mark for any ambitious young brigand, and a fast candidate for purgatory. The whole command had fallen back some twenty paces, in confusion, and were firing scattering shots at random from trees, brush, or any cover available.[41]

After a sharp skirmish with Captain Ford's troopers, the Mexicans retreated, and a messenger arrived at the rear of the train with orders for Company K to ride to the front and pursue them. This they did, with sabers drawn, and Private Cargill wrote that it took some time since the train was at least three miles long from front to rear. He also said the train stretched even farther when the wagons had to negotiate some difficult terrain. The soldiers guarding the wagons often had to help push the wagons up the sandy hills. Captain McReynold's company now remained at the head of the train and Ford's took position in the rear as the end of the train passed by. Looking back on it, Moseley reflected on his good luck in his first fight that day.

> With fatal rapidity the enemy practiced upon us. That hedge was a long blazing sheet of smoke and flames; bullets flew thick, close, and fast. Several men had fallen, killed, and wounded, and yet there was no order to charge, or flank them. Our commander raved and swore like a madman and seemed to have lost all command over himself. Yet was brave as the bravest, as had been proven on many a well-fought field. It is a trying thing to stand still and be shot at (especially for the first time) without returning fire or moving about

---

[41] Moseley, 3.

in some active way. [42]

They had only proceeded about half a mile further when a body of Mexicans fired at them from some high ground along the road. Private Samuel P. Cargill explained that their fire:

> -did not hit anyone. Captain McReynolds ordered us to charge upon them; we wheeled our horses into the chaparral and ascended the mountain. The enemy again fled, and by the time we had ascended the mountain, they were out of sight.
> We had several attacks during the day, and at five o'clock in the evening, we were attacked in front and rear at the same time. In front, from the remains of an old fort, by some 200 to 300 Mexicans. The infantry were then in front, who received the fire very coolly, and returned it promptly. The enemy, after three rounds, fled, leaving 20 of their dead; our loss was two killed and six or eight wounded. Our company was in the rear, having changed places with Captain Ford's company. We were ordered to charge upon the enemy who attacked the rear which was done in good style. [43]

After several exhausting and stressful days of attack, defense and pursuit, the convoy had managed to crawl forward until it halted six miles short of the National Bridge near the Village of *Paso de Obejas*. There was a bridge spanning a stream of good water there. Cargill wrote that they had been attacked in a similar manner ten or fifteen times, as the wagons and mule teams managed to trudge another twenty miles further. Colonel McIntosh had lost between twenty and twenty-four wagons and twenty-five men killed and wounded along with an unknown number of horses and mules up to that point. [44] Sadly, Company K and the State of Michigan suffered their first combat casualties of the war during one of these smaller fights. Private

---

[42] Moseley, 3.

[43] Ibid.

[44] L. R. Hamersley, "General Lane," 491.

Cornelius N. Combs of Tecumseh was killed in action and two others were wounded, one of them likely Private William B. Seymour of Tecumseh.[45] Private Combs was hastily buried along the side of the road by his friends, which was sadly the fate of hundreds of young Americans in that war, and those dead service members are still out there. A shallow pit was dug in the sand, far enough off the road that his grave would not be desecrated by hundreds of wagons in the days to come. Perhaps his cap was laid over his face or he was wrapped in his blanket, and then a friend spoke a few words, followed by the quiet, and the sound of the shovels, borrowed from a nearby wagon. His weapons, equipment, horse, saddle and tack would have been given to another who had had his horse shot out from under him. Within a few weeks' time, they were all well aware that his grave, along with hundreds of others along the road, would likely be looted in the darkness as locals searched for blankets, shoes and clothing. It would be a blessing if it became unrecognizable quickly, so that it would be left alone.

Questioning local civilians confirmed 1500 to 2000 Mexican irregulars awaiting them at the National Bridge. Colonel McIntosh knew his exhausted force could not storm the position without artillery support, so the convoy planned to remain where it was until someone came forward to support them. At 11PM on the night of June 6, McIntosh sent messengers back with the news of their situation and Private Cargill was one of the messengers. After making his report at Vera Cruz, he must have been relieved to hear that shortly thereafter another column of troops and wagons was already forming and would soon head up the road to rescue and reinforce his friends he had left behind. He wouldn't have to try to make his way back up the road alone.

Three days later, on June 9, the date of Cargill's letter home, he was headed out again, riding out with two pieces of artillery and five more companies of infantry under Brigadier General George Cadwalader. Cargill estimated there were 500 men with mountain howitzers, and he believed that:

> We shall now be able to cut our way through ten

---

[45] Smith, 33.

times the number of our force. Should the enemy make an appearance on the bridge, they will have a very good time. Our men have had just enough to banish their fear of powder and balls. All they want now is just to get a chance at the cowardly Mexicans.[46]

He finished his letter by mentioning the name of his friend Private Charles Tower, asking his brother to let the Tower family know that Charles was doing well so far.

Private Cargill and the men of Company K would "get their chance at the Mexicans" shortly thereafter. General Cadwalader reached the stalled wagon train on June 10, 1847. The next day they moved out together as one command, and when they reached the National Bridge on June 11, General Cadwalader was not about to let the Mexicans stop him. He had Captain Gaither's Company C of the 3rd Dragoons with him as part of his relief force, and once he arrived it appears that he kept both convoys closed up together and moved as one body. This meant that for the first time, the four companies of the 3rd Dragoons were for the most part, operating together. They would have moved cautiously up the road; they did not know what was ahead of them as they had never been to the bridge before and did not know the terrain. Cargill wrote in a follow up letter from Puebla that:

> Our train marched from that place to the National Bridge, at which place the enemy had a fort, but no cannon. On each side of the bridge is, not exactly a mountain, but a high and long hill: one side is a fort, very difficult of access.[47]

The area has changed little since that time, and the long curving stone bridge is still there. The Americans would not have seen it as they left the bluff above and started a gentle winding descent along the road until it leveled off at the same height as the bridge. Cargill described the area on the near side of the bridge as a "pass" because

---

46 Cargill, *Detroit Democratic Free Press*, July 8, 1847.

47 Cargill, *Detroit Democratic Free Press*, December 22, 1847.

from his perspective the bridge appeared to go between the two ridges, although they are not parallel. The bridge curves to the left as it crosses the river, and below it the river curves to the right just after it flows under, from left to right forming a sideways X. The "high, long hill" on the near side of the bridge looked down on and guarded the approach on the left and was topped by a small stone fort. On the far side of the river, the long high hill began at the far end of the bridge and followed the curve of the river to the right. This created an amphitheater for the enemy to fire down onto the entire length of the bridge from the safety of the other side of the river. Cargill remembered that:

> The train (which arrived here just at dark) moved along slow and cautious, expecting at every moment a volley from the enemy. I had just entered the pass when they, the Mexicans, 'let flicker.' Our men returned fire but did no execution, as the enemy lay behind their breastworks on the hill and in the fort.[48]

The Mexicans were also firing from the other side of the river, and so it must have seemed to the men of Company K that they were taking fire from every direction except behind them. In addition,

> -a force of the enemy from the bushes attacked our troops who were left in the road to guard the train, and now there was a general conflict, amid the darkness of night-the two opposing forces scarcely distinguishing their own men from their enemies.[49]

As they returned fire at the flickering lights above them along the high ground, Cargill watched the battle unfold on the near side of the bridge to his left in the darkness.

> General Cadwalader, who commanded the division, ordered a part of his men to follow him up

---

[48] Cargill, December 22, 1847.

[49] Ibid.

the hill and charge on the enemy, which they did, and soon routed the Mexicans from behind their breastworks. Our troops finally succeeded in putting the enemy to flight, except those who were stationed in the fort. The train now moved forward, regardless of the fire, which was kept up from the fort, and finally succeeded in crossing the bridge, and having proceeded a short distance from the bridge, encamped for the night. In the morning our men awoke expecting to have another fight; but the enemy had during the night evacuated the fort, and therefore saved the Yankees the trouble of taking it from them.[50]

In his report afterward, Cadwalader said that the bridge was indeed barricaded at the far end. His plan was to have the section of two mountain howitzers brought up to destroy the barricade and then rush the bridge with a combined force of infantry and cavalry in the dark. This he did, although at a cost of thirty-two officers and men killed and wounded in just that fight. Which two companies of the 3rd Dragoons were involved in rushing the barricade in the dark is unknown. Sadly, he could give no count of the losses from the teamsters and civilians moving with the train. [51] The "printers" among the Michigan men, would have been eager to get their story out, but it seems only Private Cargill, who was not one of them, succeeded. It was becoming apparent to the newsmen of Company K that it was very rare for an enlisted man to get a letter out to the folks back home when they were part of Scott's Army. They had not accounted for this reality. Private Cargill only managed it because he was physically at Vera Cruz.

General Cadwalader sent the wounded from both trains back to Vera Cruz on June 13 in a separate wagon train guarded by two companies of the Voltiguers Regiment after the fight at the bridge.[52] A gentleman who went by the initials V.R.M., was living in Vera Cruz,

[50] Ibid.
[51] Hamersley, 491.
[52] NARA Colonel Wilson's letter regarding Cadwalader June 10, 1847.

32

and was able to get a letter posted back to the United States. He listened to the reports and rumors of the fight at the bridge and wrote of McReynolds in an article for another paper stating, "that he had a victorious battle under General Cadwalader, at the National Bridge, is certain."[53] Four of the five companies of the second battalion of the 3rd Dragoons participated in that fight at some point. Only Captain Duff's Company E had not yet arrived at Vera Cruz.[54] After recovering the wounded and disposing of captured equipment, the wagon train moved on. The disposition of the dead near the bridge is unknown, but they were likely buried near where they fell. Fifty-seven men and an unknown number of teamsters had been killed and wounded so far, and they still had a long way to travel before they got to Puebla.

The convoy moved on to *Plan del Rio,* where it camped on June 13, 1847. They were all well aware that they were near the old battlefield at Cerro Gordo, which had been fought two months earlier in mid-April. It was a natural ambush point, where the road meandered through a series of easily defended hills, but safely out of the yellow fever zone. Because of the likelihood of the enemy occupying the major fortifications still in existence on the old battlefield, a cavalry reconnaissance was sent forward. No Mexican forces were found waiting for them, so they moved on. The wagon train moved through the old battlefield on June 14, past the village of Cerro Gordo and camped on the other side. General Cadwalader's column finally reached Jalapa on June 15 and remained there for three days. Colonel Thomas Childs of the 1st Artillery Regiment had been occupying the city since late April with four companies of the 2nd Dragoons, and the First Regiment of Artillery (acting as infantry.) There was another detachment further up the road in the town of Perote. That garrison was made up of the 2nd Regiment of Pennsylvania Volunteers under Colonel Francis M. Wynkoop and Company C. of the Mounted Rifles, under Captain Samuel H. Walker. This unit, unlike most of the Mounted Rifle Regiment that fought as infantry, actually had horses.[55] In both cases, the infantry was used to protect the supply depots and

---

[53] Correspondent of the Daily Detroit Advertiser as reprinted by the Oakland Gazette, Pontiac, July 31, 1847.

[54] NARA Colonel Wilson's reference to Duff arriving, June 25, 1847.

[55] Hamersley, 493.

33

hospitals in the towns and the cavalry patrolled the road to keep it clear for American use. Per instructions that had just arrived from General Scott, Colonel Childs was in the process of evacuating the hospitals and preparing his men to move to the garrison closer to Puebla by heading to Perote. General Cadwalader took Colonel Child's garrison with him when he left, creating an even larger wagon train when it left the city.

They departed Jalapa on June 18, and as they traveled up the road they were continually shot at by individuals or small groups from a distance. There was another sharp fight with Mexican forces up the road at a mountain pass called *La Hoya* near the village of *Las Vigas* on June 19, shortly before reaching the city of Perote. Colonel Wynkoop, commanding the garrison of the city of Perote, assisted in getting them in safely.

> [Colonel Wynkoop] learned that a force of fifteen hundred [Mexican] men were stationed at La Hoya, with the determination of attacking General Cadwalader and train.[56]

Cargill said that the train was halted and,

> A body of our troops sent up the mountains, each side of the road, and two mountain howitzers sent forward in the road. Soon the "white coats[57]" [Mexicans] were observed on the side of the mountain. A few rockets from our howitzers were sent whizzing among them, which scattered them in all directions. Many of them were killed by our troops stationed on the top of the mountain.[58]

---

[56] Ibid, 493.

[57] Only the 11th Regiment of Mexican Infantry was specifically issued white jackets, however, many of the local national guard or "Activo" units were issued white cotton summer uniforms. Regular cavalry normally had blue or green uniforms, and one in red. So, this was infantry not cavalry. Probably local militia.

[58] Cargill, December 22, 1847.

The troops he thought were stationed "on top of the mountain" were actually Company C of the Mounted Rifles under Captain Walker and five companies of Wynkoop's 2nd Pennsylvania Regiment that had set out from Perote to spoil the Mexican ambush. Wynkoop attacked the Mexican position from the rear as they sat waiting in ambush to stop the slow-moving convoy. General Cadwalader, meanwhile, had sent six companies forward up the hills on either side of the pass to clear it from the front. The outcome was inevitable, the Mexican forces, trapped in between, fled, and the pass was open.

On the other side of the pass was the village of Las Vigas. Wynkoop's men had cleared some Mexican rear-guard skirmishers from it earlier in the day and after the fight at La Hoya the entire wagon train came down from the mountain pass and began moving through it as darkness fell.[59] Evidently rifle fire had set some of the buildings on fire, and the scene was quite surreal. The slow-moving wagons and men were illuminated by the glowing fires of the burning village on both sides of the road as they rumbled through to the safety of Perote.

The garrison at Perote was approximately two thousand men when General Cadwalader and Colonel Childs arrived, and he noted that about half of them were in the hospital, located at the large fortress, on a hill outside of town, which was often referred to as the "castle." The convoy remained at Perote for approximately two weeks after receiving orders on June 23, to wait for General Gideon J. Pillow's column which was several days behind them and trying to catch up. Few among Cadwalader's men were upset with the order to stay where they were a few more days. General Pillow arrived on July 1, and then the much larger convoy set off for Puebla from Perote on July 2, 1847. It is not known if Lieutenant Colonel Moore and his small battalion headquarters traveled with them by then, but it is likely. It is also not known when the other half of Company G arrived with horses, but they were likely part of Pillow's column as well. The size of the immense convoy did not intimidate the Mexican forces in the area. The Americans were ambushed at 2 o'clock in the morning on July 6 while still two days out from their destination. Hearing this, General

---

[59] Hamersley, 493.

35

Scott sent the remaining companies of Colonel William S. Harney's 2nd Dragoons out from Puebla to escort General Cadwalader, General Pillow, Colonel Childs, Colonel McIntosh, and the mass of soldiers and wagons, the rest of the way to safety. Forty-eight hours later, the weary five-mile convoy had rolled into the city, where Scott's Army was waiting for them, on July 8, 1847.[60]

Until General Scott moved onward toward Mexico City in August, the dragoons performed scouting and convoy security missions in the vicinity of Puebla, coming to the assistance of one more major convoy headed for General Scott's Army, that of General Franklin Pierce, which ran the same gauntlet to get to him. In another letter written afterward, Private Cargill remembered that "the whole time the army were drilling and preparing for the march to the City of Mexico."[61]

It was at this time that Scott created a personal escort of two companies of dragoons, Captain Kearney's Company F of the 1st Dragoons and Captain McReynold's Company K of the 3rd Dragoons. Since Captain Phil Kearny was from the Regular Army and had date-of-rank, he was the overall commander of the escort. "From these two companies, General Scott chose ten men for his personal bodyguard, but only Private N. A. L. Simonds, of Raisin Township, Lenawee County, has been identified as a member of this select group from Company K."[62]

Despite being away from the coast, the" Yellow Fever" season was in full swing, and soldiers began to fall ill across the entire army. Large hospitals were set up and surgeons and attendants from different units were concentrated in those places, primarily in major cities, where housing and water supplies were available. The hospitals filled rapidly, and more men were detached from their units to act as hospital stewards. The sick were sent back to their units if they improved, and those who didn't were slowly evacuated to the hospitals in Perote and Vera Cruz until they either recovered and went back on duty or were evacuated even further back to hospitals at New Orleans. During the month of July, the following Michigan men of K

---

[60] Cargill, December 22, 1847.

[61] Ibid.

[62] Richard Illenden Bonner, ed., Memoirs of Lenawee County Michigan. (Madison, 1909), I, 642.

36

Company were left behind or evacuated back to the hospital in Perote, Mexico:

Private Gilbert Ball, a twenty-one-year-old farmer from Detroit who had been detached as a teamster in June, ended up in the hospital and died there later, on August 6, 1847. His twin brother Alden would be discharged the following month from a hospital in Mexico City.

Private John L. Leavenworth, a twenty-two-year-old from Detroit, was left there on July 2, 1847. He may have been evacuated to Vera Cruz. He died "In Hospital" on September 22, 1847.

Private Herman Harris, a twenty-one-year-old from Tecumseh, was left there on July 3, 1847.

Private John Hill, a thirty-five-year-old from Tecumseh, was also left there on July 3, 1847.

Private John Sly, a twenty-three-year-old who had been a waiter from England, died on the 4th of July 1847 in the hospital at Perote.

Private Milton A. Wood, a nineteen-year-old from Tecumseh, died of disease on July 3, 1847, in Perote.

Although not mentioned in pension records, Cargill mentions in his letter home that his friend Charles Tower had been left in Puebla until he recovered.[63] There are several descriptions of where all of the dead were buried at Perote and are still lying in unmarked graves. A long ditch was dug on the south side of the dry moat outside the castle wall, to the left of the main gate. All of the Americans who died at Perote were buried there in long rows, and there they remain.

---

[63] Cargill, December 22, 1847.

CHAPTER THREE

To the Halls of Montezuma.

A guidon was captured by our men bearing the
inscription 'No quarter to the cursed Yankees.'
-D. H. Hill[64]

The march to Mexico City from Puebla began on August 7, 1847, When General Twigg's Division left, followed by the other divisions of the army, a day apart from one another.[65] General Scott and his escort left the city traveling between the divisions on August 8. It was not, however, until August 10th that the entire Army had left the city except for a small garrison under Colonel Childs which was left to protect the estimated 2,500 men inhabiting the makeshift hospitals there. Captain Ford and Company D were left behind to provide patrolling and scouting services as part of that garrison. The Army had been reorganized by General Scott and was "divided into four divisions, with each appropriate allowance of dragoons and light batteries."[66] Sometime during August 8, Company K left Puebla with General Scott. Lieutenant Moseley described the spectacle of his (General Worth's) Division leaving the city.

As our division debouched from the gates of the

---

[64] Hughes, Nathanial C. & Johnson, Timothy D. ed., "A Fighter from Way Back: The Mexican War Diary of Lieutenant Daniel Harvey Hill, 4th Artillery, U.S.A.," Kent State University Press, Kent, Ohio, 2002. 108

[65] Cargill gives the date as the 5th.

[66] Moseley, 3.

city, and fell into the regular route-step of march, infantry, cavalry and artillery, arrayed in all the panoply of war, with burnished arms, bright uniforms and glittering decorations; the bands playing stirring martial airs, the colors of the different regiments fluttering in the tremulous breeze, as they were waved in the passing salute; the rattle of drum and the coarse blast of the cavalry bugle; all filled the heart of the young soldier with proud and daring emotions.[67]

Despite the fanfare, things were not going well for many of the Michigan men in Company K. The environment had finally caught up with them, and they were becoming sick in large numbers. The rest and recovery time they received at Puebla seemed to be making things worse. The longer they remained, the more men became sick. They left over twenty-five percent of the company behind when they marched out of the city. Those twenty-eight Michigan men from Company K left behind in hospital on August 7, 1847, were:

Private Andrew Albro, a twenty-one-year-old blacksmith from Detroit.

Private Steven Armstrong, a twenty-one-year-old blacksmith from Detroit.

Private Daniel Axford, a twenty-nine-year-old Canadian farmer that crossed the river to enlist in Detroit.

Private Ira W. Barlow, a twenty-year-old miller from Detroit.

Private Joshua Beaman, a.k.a. Bedman was a twenty-five-year-old farmer from Tecumseh, Michigan.

Private Clement Beniteau, a.k.a. Boniteau, a twenty-six-year-old clerk from Detroit or perhaps Canada.

Private William H. Caleb, a nineteen-year-old farmer from Tecumseh.

Private Isaac Carey, a twenty-six-year-old farmer from Detroit.

Private Henry B. Cornwell, a twenty-two-year-old clerk from Detroit.

Private George C. Daily, a twenty-eight-year-old shoemaker from Tecumseh.

---

[67] Moseley, 3.

Private Nathaniel B. Hall, a twenty-three-year-old farmer from Detroit.

Private Ambrose Hickox from Ann Arbor, age and occupation unknown.

Private Alpheus S. Holloway, a twenty-nine-year-old carpenter from Tecumseh.

Private Edwin Howell, a twenty-four-year-old farmer from Detroit.

Private Joel M. Jackson, a twenty-one-year-old farmer from Detroit.

Musician James Johnston, a thirty-four-year-old carpenter from Detroit.

Corporal John W. Leake, a twenty-eight-year-old merchant from Tecumseh.

Private Alonzo Lurvey, a twenty-one -year-old cooper from Detroit.

Private Robert G. McIntire, a twenty-three-year-old mason from Canada.

Private Peter Nelson, a thirty-one-year-old painter from Detroit.

Private William C. Payne, a twenty-two-year-old farmer from Cleveland, Ohio.

Private John Riddle, a twenty-eight-year-old farmer from Tecumseh.

Private Alonzo Seeley, a twenty-two-year-old cooper from Detroit.

Private William B. Seymour, a twenty-two-year-old farmer from Tecumseh.

Private Stephen Smith, a twenty-six-year-old sailor from England.

Private Senge Streeter, a volunteer from Tecumseh, age and occupation unknown.

Private Samuel W. Thorpe, a twenty-five-year-old farmer from Tecumseh.

Private Charles H. Tower, a twenty-nine-year-old clerk from Detroit.

As each division moved out on the road toward Mexico City, they moved to the next site with good water. About one day apart, they came to the first bivouac site, at the village of *San Martin*. The first evening was disrupted by an approaching column of Mexican cavalry. Lieutenant Moseley recounted,

We were just beginning, after the approved style, to discuss the contents of our haversacks, with now and then some delicate morsels of foraged dainties, and beginning to feel ourselves at home, even in the little hamlet of San Martin, when suddenly, and with startling distinctness, the ominous long roll of the drum electrifies the whole mass. To Arms! To Arms! Fall in! Fall in! Prepare to mount! Cannoneers to your pieces! Away go the haversacks, tin-cups, mess-pans, and all the paraphernalia of the bivouac.[68]

The entire division was arrayed for battle for a certain period of time, until an aide rode up and whispered something to the general, who ordered everyone back to their camps. The dust cloud of approaching Mexicans had actually been a group of messengers from another division bringing dispatches to General Worth. Lieutenant D. H. Hill of the 4th Artillery, acting as infantry, who would become a Confederate general in the later war, was at San Martin when Scott and Company K rode by and mentioned it in his journal. However, he detested the volunteers, being a regular army officer himself, and so he wrote, "Genl. Scott and staff passed us today escorted by Capt. Kearny's Company.[69]" He would not recognize McReynolds at all, just Kearny, the regular army officer.

Colonel Harney of the 2nd Dragoons described the daily operations of his units between Puebla and Mexico City this way,

> The dragoons, from the commencement of the march from Puebla, have been engaged in the most active and laborious service. These duties have been the more arduous in consequence of the small force of cavalry, compared with the other arms of the service. Small parties being constantly engaged in reconnoitering and on picket guards, the utmost vigilance and precaution have been required to

---

[68] Moseley, 4.

[69] Hughes, "A Fighter from Way Back." 106.

prevent surprise and disaster.[70]

As the army ascended into the mountains that separated them from the Valley of Mexico, the weather turned cold, and those who had not jettisoned their great-coats and blankets were glad they had kept them. It is worth mentioning that all who wrote about the march and the days and events around Mexico City talked about the weather. Very specifically, it was hot and humid every day and as the sun set it would cloud over and begin to rain. Every night they slept in the rain, sometimes it rained all night until the sun rose again. The number of men becoming sick was steadily growing. The forest had turned into tall pines and the higher they went into the mountains, the cooler it became. As the march continued, the casualties also continued.

On August 8, 1847, Private Robert R. Longstreet, a teacher from Detroit, was "lost" on the march between Puebla and St. Martin. The details surrounding this event are unknown, but he was never seen again. He was considered "missing in action" and presumed dead.

On August 10, 1847, there was a fight during the march at Nuetin Nueva and Private William Gibson, a carpenter from Detroit, was killed in action. He was most likely buried on the spot by his brother Isaac.

Scott's Army descended from the mountain pass as a single mass, coming to a relatively open area on the western slopes, and the entire valley of Mexico came into view as the men came around the curve. This was a significant emotional event mentioned by almost every primary source that was part of this army. They remarked on being able to see the domes of the cathedrals of the capitol city above the shimmering lakes that lay out around it in the distance. Around the city far off to the north, west and south, stood purple mountain ranges in the haze, all visible from the same spot on the road, as they emerged from the dense pine forest. Moseley commented that the army looked like a large glittering serpent, snaking its way down the mountain to the valley floor ahead of him.[71] The army came upon the first sign of Mexican defenses near the city of *Ayotla*, on August 11, where Santa

---

[70] "The Battle of Churubusco-Report of Colonel Harney" Report of the Secretary of War, 1847, (Washington, 1848), p. 347.

[71] Moseley, 5.

42

Anna had constructed a line of fortifications defending the eastern approaches to the city. Scott had no intention of accepting Santa Anna's invitation by attacking these defenses. As the American army left the mountains into the Valley of Mexico, Sam Cargill described how he saw it.

> Our march was uninterrupted until we arrived at Ayotla, about 35 miles from the city. At a short distance from this place, the enemy had constructed a fortification of great strength, on a hill (the Mexicans called it Penon) situated as to command the road and some distance on all sides. General Scott, after having reconnoitered the fortification, determined to march by another road, which the Mexicans considered impossible for a train to pass; but they had not yet found out the Yankees yet. The road, although it was for some distance very narrow and very rocky, the Yankees did pass, and on the 18th of August arrived at San Augustin, a small place 12 miles from the city.[72]

General Scott sent scouts out in several directions to find a way around. It was Lieutenant Moseley's Company G of the 3rd Dragoons, which led the reconnaissance on August 14, 1847.

> With a detachment of men from my troop and a battalion of infantry, all under the command of Lieutenant Colonel Duncan, we made a reconnaissance of the road on the southern shores of Lake Chalco and Xochimilcan, with a view to its practicality for infantry, cavalry and artillery.[73]

The Army concentrated at San Augustin, and left detachments of cavalry to make sure the road behind them was left open but only to Puebla. It was a bold move, Scott had cut himself off from Vera Cruz, his entire available force was now isolated and alone. To people back

---

[72] Cargill, December 22, 1847.

[73] Moseley, 5.

in the United States, it was as if they had disappeared, swallowed up by Mexico itself.

For several days after arriving near the city, General Scott sent teams of engineers out to conduct reconnaissance of the roads so that he could further develop his plans to capture it. On August 15, Lieutenant Schuyler Hamilton, one of General Scott's aides de camp, rode out with two companies of the 3rd dragoons, one of them recognized as being Company K of the 3rd by D. H. Hill, who wrote about the mission in his journal. They were also supported by a small unit of cavalry called "Dominguez's Spy Company," and a small body of infantry. Hill was not complementary of what he heard about Captain McReynolds or the 3rd dragoons in his journal entry, convinced that the "volunteers" almost got his acquaintance killed.

> The cavalry was attacked when some distance from the infantry, by a large body of lancers. "Captain McReynolds and another 3rd dragoon officer ran. Lt. Hamilton was dangerously, and two dragoons slightly wounded and several of the Mexicans of our party killed. The enemy was repulsed, leaving several of their number dead on the ground. A guidon was captured by our men bearing the inscription 'No quarter to the cursed Yankees.' These guerillas were organized by General Sala.[74]

Hill was not there and did not observe any of the events himself, and Hamilton was out looking for trouble, and he found it. Again, Hill regularly made disparaging remarks about volunteers to anyone who would listen, because he considered them a nuisance, and a threat to his own promotion and career. On August 18, General Scott sent Captain Robert E. Lee and Lieutenant P.G.T. Beauregard around the southwestern edge of the road network. Captain Kearny's dragoons (and therefore we can presume Captain McReynold's Michigan boys as well) provided escort and security once again. Cargill went out with them and described the moment that their reconnaissance came around the foot of the hill known as *Zacatepec*, at the edge of the

---

[74] Hughes, 108.

Pedregal, a square mile of rocky, crevice filled lava field south of the city.

> We had traveled about 2 miles and were passing around a large hill when we were fired upon by a party of Mexicans who were concealed among the crags and rocks which extended for several miles around, over which it was impossible for horsemen to pass.

One of Kearny's officers, Lieutenant Richard S. Ewell, advanced with his troop of cavalry to push back some Mexican lancers by long range fire. Cargill noted that,

> The order was given to dismount and charge on foot, which was done, killing 10 of their number and taking 6 prisoners, and putting the rest to flight. At this time one of the party observed about one and a half miles distant something which he thought to be a fortification. By the aid of a glass, it was found to be one. We then marched back to San Augustin.[75]

It was this critical mission that convinced General Scott to circle to the southwest and bypass Santa Anna's positions on the eastern side of the city. Clearly General Scott was using his mounted assets wisely by detaching his cavalry escort to keep his most valuable assets, his engineers, protected.[76]

The fortification they had seen was at *Padierna,* near a place called *Contreras,* which gave the battle its name the next day. Mexican General Valencia had taken his troops out from the city defenses and redeployed them perpendicular across the path of the incoming Americans. Although it was a favorable defensive position where he dug in his estimated 6,000 men and his twenty-three pieces of artillery, Santa Anna and General Scott both saw the problem with this immediately. Valencia was too far from the other Mexican defenses

---

[75] Cargill, December 22, 1847.

[76] Timothy D. Johnson, *A Gallant Little Army: The Mexico City Campaign,* University Pess of Kansas, 2007. 160

and could be cut off and destroyed before other Mexican troops could march out to support them. Despite being ordered back, Valencia stayed where he was, and throughout the night of August 18, 1847, General Scott made plans to cut Valencia off and destroy him the next day.

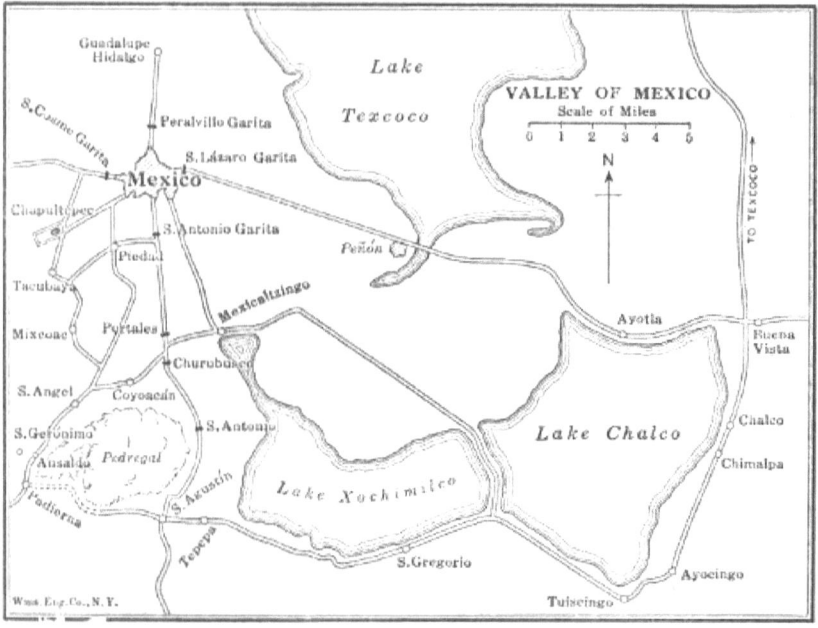

Battlefields around Mexico City (The War with Mexico, Vol. II, Justin H. Smith, 1919 .)

On the morning of August 19, the men of Company K watched General Pillow and General Quitman head west down the road and begin the process of widening and improving it under the watchful eyes of the engineers. Scott had tasked Captain Lee with finding a way through the Pedregal, at least for the artillery, even if it had to be pulled over the rocks by hand, and Lee, one of the best engineers, who had created a path to victory at Cerro Gordo, succeeded in finding another one during the night. This way Scott could put a force between Valencia and Santa Anna and destroy the Mexicans in the fortification before General Santa Anna could save them. The battle commenced as the Mexicans attempted to stop construction on the road and General Valencia opened fire with his long-range guns and sent swarms of light infantry skirmishers forward into the Pedregal to pick off the Americans. Pillow's response was to throw all of his troops into the fight and request more. This brought General Scott to the top of the hill called Zacatepec very quickly thereafter, and his escort of the Michigan men as well. Scott had not wanted to commit to a battle until his road was finished. While Scott and Pillow determined what to do next, Company K was positioned at the foot of the front of the hill, where they had dismounted the previous day, and had a front row seat to watch the two sides battle it out. They had no role to play unless the Mexicans ran, or the Americans did. They spent the day dodging Mexican artillery rounds and must have relocated several times. Another eyewitness that day, Frederick Zeh, an artilleryman, passed their location in front of Zacatepec that morning and stated, "Here we passed several dead horses, The head of one of the animals had been mangled by a shot, and that of another had been drilled straight through by a big cannonball, hollowed out, so to speak."[77] No casualties were reported at Contreras, but they may have been hidden by the later activities of the day.

The back and forth fighting through the Pedregal and artillery duel between Mexican and American artillery continued until dark.[78] It seemed as if the Mexicans had won, at least General Valencia certainly thought so, and despite more orders from Santa Anna, he still did not move. Valencia expected the Americans to attack his fortifications and be defeated the next morning. The Americans did attack, just not from the direction he thought they would. All night long, units of American infantry had infiltrated through the Pedregal down the trail constructed by Captain Lee and lined up in position beside and behind the Mexican fort. At dawn the American forces opened fire and charged the position from behind, taking it in less than thirty minutes. Valencia's force could not react quickly enough and crumbled, with soldiers running north toward Mexico City, and into American hands, or south into the countryside. It was considered a great victory, but the day was not over.

---

[77] Orr, William J., & Robert Ryal Miller, ed. "Frederick Zeh: An Immigrant in the Mexican War." Texas, A&M University Press, College Station. 1995.

[78] Cargill, December 22, 1847.

CHAPTER FOUR

Churubusco and the Charge of Company K.

Bravery may be carried too far.
-General Winfield Scott[79]

By following the whereabouts of General Scott, we can track the location of Company K. for at least a portion of the day on August 20, 1847. It had rained hard all night, and the American assault on General Valencia began at first light. Scott had returned to San Augustin the previous night when the rain started and had been up for most of the night planning for what he would do once the morning battle had been decided. Private Isaac Gibson of Company K remembered that soggy morning quite clearly and wrote about it later in life.

> The Grey Horse Squadron (General Scott's Bodyguard) commanded by Captain Andrew T. McReynolds and the lamented Phil Kearny, were early in the saddle on that day, in fact most of us were in the saddle all night. Valises, blankets, overcoats, and haversacks were laid aside in the company wagons at San Augustin, and we were ready for action.[80]

After the early morning battle of Contreras had turned into a

---

[79] O. L Ray "Reminiscence of Scott's Campaign." *The Vedette*, Volume 1. (April 15, 1880): 13.

[80] Isaac Gibson "Phil Kearny's Charge: An interesting letter from Michigan." *The Vedette*, Volume 1. (April 15, 1880): 12.

Mexican retreat, General Scott and his staff mounted their horses and followed the left wing of the American Army along the southern causeway as it advanced north toward the village of Coyocan. They returned to the battlefield they had left the night before and with the dawn, the destruction was evident everywhere. Following the track across the battlefield to the main road at Contreras and turning north, they picked up other commanders along the way, most notably General Pillow. General Scott stopped from time to time to speak with the victorious units as they marched along past the destroyed remains of General Valencia's forces, toward whatever was awaiting them at the front of the American army. As the sun rose, Captain Kearny and McReynold's men rode along in front and rear providing escort. Gibson recalled,

> Marching over the dead and wounded on the field of Contreras, we reached the vicinity of Churubusco, and, by the side of our great commander Scott, dismounted to allow a few moments of rest to our horses. Soon, General Scott, in a quiet tone, gave the order, 'to horse Gentlemen;' every man was in the saddle in an instant; headquarters bugler sounded the call to arms the buglers of other commands repeated the call, and the 'long roll' of the infantry was heard from every brigade-all were in motion. As hours passed, the squadron was moved from place to place, where it would do the most good.[81]

The army had come upon a new Mexican battle position, and the lead elements of General Pillow's and General Twigg's forces were already skirmishing with Mexicans in the cornfields that lay in front of it. Based upon reports of the enemy fortifying the North side of the river at *Churubusco*, he sent Captain Kearny (and therefore we can presume Captain McReynolds) forward to explore the best approach routes to attempt a flanking move on the left, west of the *San Antonio* Road. He ordered General Pillow to support them with General Cadwalader's Brigade from his division and they headed off toward

---

[81] Isaac Gibson, "Reminiscence," 12.

the rain drenched fields. They were still on that mission out to the west of the fight about an hour later, as General Scott realized no friendly forces were behind him and he was required to move up to a dangerous position just behind General Twigg's Division for protection, specifically because he had no escort. [82]

The rest of the 3rd Dragoon companies remained with Worth's Division and followed up the assault on the Mexican lines on the right-hand side of the battlefield at the village of San Antonio that morning. Lieutenant Moseley wrote that Company G followed the artillery, which followed the infantry up the road toward Churubusco. Worth's Division advanced toward the bridge over the Rio Churubusco and found a large earthwork with cannon blocking the path, called a Tete-du-Pont. General Worth spent the morning bombarding and then assaulting that fortified position, as General David E. Twiggs and General Pillow's men from the other divisions came up another road on his left, fresh from the victories at Contreras. Those forces assaulted and captured a large convent to the left of the bridge, San Mateo, which was also heavily fortified. The two forces merged as the positions fell to the Americans after a very difficult and costly struggle. Scott's two squadrons returned and were present as the line of Mexican defenses finally broke at the Churubusco bridge. They had arrived just in time to follow up the breakthrough, a job for which the dragoons were specifically created, to exploit and chase down a defeated and fleeing enemy. Cargill, who was there, reported that they, Kearny and McReynold's men,

> -were ordered to charge upon the retreating enemy (the order not being given until they [the Mexicans] were halfway back to the city.) The cavalry charged at full speed, overtaking and passing by many of the fleeing Mexicans who could not escape from the road on account of a wide and deep ditch on one side and a lake on the other. [83]

This was the key moment in the history of the 3rd Dragoons, and in

---

[82] Timothy D. Johnson, *A Gallant Little Army*, 187

[83] Cargill, *Detroit Democratic Free Press*, December 22, 1847.

the careers of its officers. This was the moment for a glorious charge, sabers flashing, victory certain, accolades, and promotions awaited them at the outcome of such a moment, and they were keenly aware of it. There were no medals to be awarded back then, and only a daring deed, witnessed and approved of by others tested a man's bravery, his military bearing, professionalism in his art, and his honor as a soldier. This was the moment, and they could not hesitate. It was four o'clock when the charge began, and although the fighting had ended at the bridge near Churubusco, O. L. Ray, another soldier who was there recalled that lighter guns of the flying artillery (actually probably the two mountain howitzers of Lieutenant Reno's battery) were still firing down the causeway three quarters of a mile to the north. The narrow stone causeway was jammed with fleeing Mexican infantry. They were not just clogging the road, they were also crossing the ditches and heading back to the city across the swampy fields, swimming the irrigation ditches when they had to. Ray estimated that the entire distance from Churubusco to the well defended gate at San Antonio was four miles, although this seems extreme. He described the gate itself as being defended by two lunettes with cannons aimed straight down the causeway. He also explained how the Mexican defenders had taken the causeway apart, stone by stone, so that the road was cut with a large gap that had filled with water from the irrigation ditches nearby, creating a perfect moat.[84] It is doubtful that Kearny and McReynolds knew about the gap since no Americans had gotten to that part of the battlefield or been that close to the city itself. The opportunity to cut down the fleeing enemy was too great, and the risk seemed so low, and the chance for glory and victory never so obvious. How could they not go? It must have felt like destiny.

Lieutenant Moseley, who was supposed to be acting as General Worth's bodyguard for the day recalled,

> Just as the tete-du-pont was captured and I was pushing forward with my detachment, the causeway was blocked up with the huge Mexican ordnance wagons, one of which was on fire, threatening instant explosion. I immediately ordered Sergeant A. M.

---

[84] O. L. Ray Memoir. The Vedette, Volume 1. April 15, 1880, 12.

Kenaday, of my detachment, to dismount with some of the troopers and put out the fire and throw out the ammunition. This perilous duty, equal to storming the deadly breach, was executed in the most gallant and expeditious manner, and a calamitous disaster averted by true heroism.[85]

Moseley then charged forward following Captain Kearny and Captain McReynolds down the causeway with his men of Company G. Mounted officers from different commands, aides of some of the generals, anyone with a horse, seemed to be pulled into the charge. Major Frederick Mills, one of the officers of the 15th U. S. Infantry Regiment who had watched the charge go in from off to the left of the causeway, took off after them, his horse racing to catch up. Some say his horse had a mind of its own that day and he had completely lost control of it.

O. L. Ray recalled the incident with Sergeant Kenady and the artillery wagon as well and said that the halt to get past was frustrating because it allowed the enemy to fall back at least a mile toward the gate before the attack resumed. It is unclear if they followed the same route as Captains Kearny and McReynolds to the causeway, and the distances given are questionable. He observed that,

This small, mounted band, now forming in sets of four or five, started in full pursuit. They were closely followed for a time by Colonel Harney, without a command. Near Portales (a hacienda on the west side of the road) they encountered a flank fire from the Mexican reserves under General Perez. Heedless of this they sped on, the way being open, and overtook the enemy about a half mile outside the city gate. The foe immediately in front outnumbered their assailants by fifty to one. [86]

---

[85] Moseley, 8. In Frederick Zeh's account, he described the cavalry charge swept past them toward the gate from right to left down the causeway.

[86] O. L. Ray, "Reminiscence." 12.

52

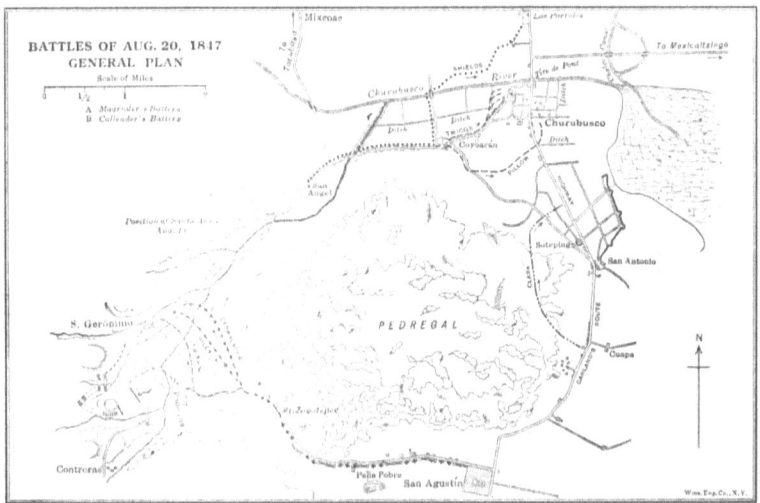

Battle of Contreras and Churubusco

(The War with Mexico, Vol. II, Justin H. Smith, 1919.)

They spurred their horses on and fought their way into the mob of enemy soldiers. It seemed foolhardy that they should attempt to make it all the way to the gate, what would it possibly accomplish? What would they do when they reached the literal end of the road? General Scott, having seen what was about to happen, and how this would likely end, sent riders forward to Colonel Harney and had his buglers blow the recall, and those who could hear the bugle calls to retreat slowed, and turned their horses back the other way, but the men in the front did not hear.

Ray counted only three or four sets of four or five horsemen each still charging forward by the time they approached the ditch and gate itself. The enemy were fleeing down into the ditch, up the other side and climbing over or going around the artillery position beyond, this allowed six riders, Captain McReynolds, Captain Kearny, Lieutenants Graham and Hamilton, Major Mills and Sergeant Kenaday to ride down into the ditch and up the other side along with the fleeing Mexicans. About a dozen horsemen of the mixed companies made it to the near side of the ditch at about the time the Mexicans stopped running. Within what must have seemed like only seconds they

53

turned, ran back to the two cannons, and began firing deadly clouds of grapeshot toward friend and foe alike. Survivors from that day witnessed Major Mills jumping his horse over the breastworks, never to be seen again.[87] He was the only one who went farther than the ditch. Captain McReynolds, Lieutenant Graham and Sergeant Kenaday were hit shortly after arriving at the ditch, but Captain Kearny seemed impervious, although he had lost his sword belt, holsters, pistols and most importantly, his horse. The tables had clearly turned, and the attackers were now attacked. Those Americans that were on the near side of the ditch attempted to extricate their leaders from the far side in the face of increasing musket and cannon fire as the Mexican infantry also turned, loaded their muskets and began firing at Americans stalled in and across the ditch. Those closest to the enemy and still mounted fell back if their horses could make it back across the water and up the steep bank to safety. A riderless horse was found for Captain Kearny. Exhausted, but still grasping his sword, he mounted the horse and was then immediately hit in the left arm either by Mexican grapeshot or a musket ball.[88]

They had a long ride ahead of them, on exhausted horses, with enemy artillery firing at them from behind as they went. Dragoons that were still mounted tried to pick up those who had their horses shot out from under them, and some of them just began to run back down the causeway. Some of the wounded were recovered by others as they fell back. It would have been extremely difficult trying to get out of the way of the deadly fire while making the most conspicuous targets, being on horseback. The minutes went by, and the distance increased until Mexican fire slackened and then stopped. The survivors of the charge, wounded or not, would have certainly been in a state of shock, especially the wounded. Those that found themselves without a wound would have certainly considered themselves lucky. General Scott's official headquarters report stated:

> The cavalry charge was headed by Captain Kearny, of the 1st dragoons, having in squadron with his own troop, that of Captain McReynolds of the 3rd –making the usual escort to

---

[87] Johnson, 190

[88] O. L Ray Memoir, The Vedette, Volume 1. April 15, 1880, 12.

general headquarters; but being early in the day detached for general service, was now under Colonel Harney's orders. The gallant captain, not hearing the recall, that had been sounded, dashed up to the San Antonio Gate, sabering, in his way, all who resisted. Of the seven officers of the squadron, Kearney lost his left arm; McReynolds and Lieutenant Lorimar Graham were both severely wounded, and Lieutenant R.S. Ewell had two horses killed under him. Major F.D. Mills, of the 15th Infantry, a volunteer in this charge, was killed at the gate.[89]

First Lieutenant Brown, who had been languishing in the hospital from disease, had managed to leave his bed with the assistance of friends and "tonics" and mounted his horse at some point during the battle. He was there when the order to charge was given, and somehow managed to stay in the saddle and make the charge with the company.[90]

A letter was written at the National Palace in Mexico City on October 29, 1847, (possibly by Lieutenant Colonel Moore) and signed simply as "A Citizen of this place." It was sent back to Lieutenant Brown's father, a retired general from the War of 1812 and later published in a local newspaper. The author explained to General Brown that,

Lt. J.T. Brown can tell you all about our terrible battles since we have entered Mexico, in which he has borne an honorable and distinguished part. In the gallant charge after the route of the enemy at Churubusco, on the 20th of August, made by the squadron of dragoons under Capt. Kearny, composed of his own and Capt. McReynolds' troops which formed the escort of the General-in-Chief, he bore a

---

[89] An excerpt of General Scott's After-Action Report. Printed in the Detroit Democratic Free Press, December 20, 1847.

[90] Andrew T. McReynolds, Pioneer Society of Michigan, Annual Meeting of 1878, pg. 433

conspicuous part. This handful of men pursued the flying Mexicans along the great southern causeway (the same by which Cortez entered the city) even to the Garita San Antonio, when Kearny lost his arm and was assisted in retiring by Brown.[91]

O. L. Ray observed the exhausted men and horses as they returned from the fateful charge and described their sad state,

As Captains McReynolds and Kearny neared headquarters, each supported between two dragoons (one of them First Lieutenant Brown of Company K who had left his sickbed) 'covered' as they were with wounds and with honors. General Scott approached and publicly complemented them for their bravery. At the same time, he hinted that 'bravery may be carried too far.'

Colonel Harney was more sympathetic. Advancing to McReynolds he said, 'Captain I would give ten thousand dollars for that wound of yours.' The captain faintly replied, 'If you will relieve me of it, you can have it for half the sum.' "[92]

Nothing is mentioned of the retreat of the men after the impetus of the charge was spent, nor of the attempts to rally the survivors and get accountability as they attempted to reform, with foaming and exhausted mounts, and empty canteens, probably somewhere near General Scott's headquarters flag. Chaos would have reigned, especially since both commanders were down, how the wounded were brought in was unclear, although it is known that both Kearny and McReynolds remained in the saddle until they got to a hospital. How the junior officers and enlisted men made it back and what state they were in is unknown. Scott wrote in his report that "both of these fine companies sustained severe losses in their rank and file also."[93] It

---

[91] Letter to the *Toledo Blade* reprinted in the *Hillsdale Whig Standard*, December 28, 1847. The letter also mentions the fact that Lieutenant Brown is severely ill from disease.

[92] Ray, 12.

[93] Excerpt from Scott's Battle report printed in the *Democratic Detroit*

would have been up to the Lieutenants, Ewell from Kearny's Company, Williams, Henry, and Brown from Company K, to find a rally point, recover the wounded and get accountability of the survivors. It is very likely that Brown went right back to the hospital as well. The eyewitness involved, Private Cargill, simply stated that "The cavalry now marched to Churubusco, remaining there that night." [94] Kearny had his arm amputated shortly thereafter, his head held during the operation by another officer, future President of the United States Franklin Pierce.

An American Dragoon officer charging during the Mexican War (Infantry Museum at Fort Benning, photo by author.)[95]

This charge was forever remembered as Company K's great moment, although it is known historically as "Kearny's Charge." Alas,

---

Free Press, December 15, 1847.

[94] Cargill, December 22, 1847.

[95] The officer in the photograph is Captain May of the 2nd Dragoon Regiment, but the uniform and horse equipage for the Michigan officers would have been the same. The photo was taken at the museum on March 3, 2025.

Michigan is not given credit. It was one of the last actions by General Scott's army that day and was clearly a gross overreach. It made no difference to the day's events. Mexican forces had stopped running and turned by then and the chances of victory even if the dragoons had reached the gate itself were non-existent. The number of dragoons in the charge, based upon troops available and the volunteers who piled on (such as the unfortunate Major Mills of the 15th U. S. Infantry Regiment) still probably did not exceed one hundred men, or one full company of sabers. On the narrow causeway, in the face of Mexican guns firing grapeshot point blank into their ranks, it is a wonder that more of them were not mowed down. They would have been slowed to no more than a canter by the mob of fleeing Mexican infantry and would have been hard pressed to move any faster on the narrow causeway. Moseley and the men from Captain Duperu's Company G appear to have turned back (except for Sergeant Kenady) when they heard the buglers blowing the recall order. Even so, he lost one killed and one wounded. In the end, there were six dead privates left at the far end of the causeway, along with their horses. Besides the officers and Sergeant Kenaday, there were three more privates badly wounded, many slightly wounded, along with all of the horses lost. That the casualties were as low as they were, was nothing short of miraculous.

The Michigan papers discussed little of the carnage and much of the glory. The citizens of Detroit read the following in the Democratic Free Press.

> It was in this charge that Captain McReynolds, of this city, received his serious wound. His troop-all Michigan boys-together with Kearny's, participating. It was undoubtedly one of the boldest, and most daring charges on record. The commanding general of the division thus speaks of the charge, and Captain McReynolds and his bold dragoons: 'Captain McReynolds' 3rd Dragoons, nobly sustained the daring movements of his squadron commander, and was wounded in his left arm. Both of these fine companies sustained severe losses in their rank and file also.
>
> In every engagement the 'Michigan Boys' have covered themselves in glory. The infantry and

dragoons from this state deserve and have received by the dispatches, great credit.[96]

The night of August 20, 1847, was rainy and black, and Captain Duperu's Company G, the Louisianians of the 3rd Dragoons, were chosen to escort the English Ambassador to General Scott to negotiate an armistice with Santa Anna.[97] Both armies were exhausted, and General Scott had been instructed to encourage peace negotiations should the opportunity present itself. There was a truce in place for several weeks after this as Santa Anna negotiated publicly for peace but privately for time, and General Scott attempted to win without another bloodbath, but this was not to be. It was clear after a few weeks that peace talks were going nowhere and that the Mexicans had been using the break in fighting to reinforce the city garrison instead, and so hostilities recommenced in the first days of September.

Meanwhile, far to the North, the headquarters and companies of the First Battalion, 3rd Dragoons, assigned to General Taylor's command conducted convoy and occupation duties. It was boring work. A reporter from the New Orleans Picayune visited the camp of the 3rd Dragoons and described his observations of Major Cass.

> Cass is silent and reserved, living entirely alone, a complete solitaire. He has a beautiful bird of the falcon species, spotlessly white, which is the only guest ever admitted to his tent. His officers say, that, except to issue orders, they scarcely ever hear the sound of his voice.[98]

In his defense, the major was the only person from Michigan with that half of the unit. With Colonel Butler there, and the headquarters staff, and only half of the companies of the regiment to command, there may not have been much for him to do. The northern theater of the war quieted down significantly after the Battle of Buena Vista,

---

[96] Ibid.

[97] Moseley, 9.

[98] *The Jackson Patriot*, quoting an article originally printed in the *New Orleans Picayune*, August 31, 1847.

which they were not present for, and there is very little record of what the unit accomplished, other than where the different companies were stationed and when.

CHAPTER FIVE

Molino del Rey and Mexico City.

*-a lone Mexican Lancer rode out from his lines in a way
that the Americans interpreted to be an invitation to
individual combat. An American sergeant accepted the
challenge by riding forward to face his opponent, and when
a trumpet sounded, the two men charged at one another at
full speed. At the last instant, the dragoon veered left,
dodged the lance, and, with the swing of his sword, almost
decapitated the Mexican. Then, with a coolness that drew
the admiration of his comrades, the sergeant grabbed the
reigns of the lancer's horse and took it back with him to his
own lines.*
-Moses Barnard, Voltigeur Regiment[99]

On August 21, 1847, the day after Churubusco, General Scott established his headquarters at a large stone complex of religious buildings known as the Bishop's Palace in the small town of Tacubaya, south of Mexico City. Representatives of the Mexican government came and requested an armistice while they debated on what to do next, negotiate with the Americans or continue the war. Knowing that his army would only get smaller until reinforcements came and not wishing to shed more blood on either side, General Scott agreed to it. The American army had been battered quite badly as well. Many in the

---

[99] Johnson, 207.

61

army disagreed with this and believed correctly that it was just giving Santa Anna time to rebuild his forces.

On September 6, 1847, after several incidents between the two antagonists, the armistice dissolved, and both forces squared off to continue the fight. General Scott decided to remove the threat of some Mexican defensive works in front of the tallest terrain feature outside the city, the Castle of *Chapultepec*, which the Americans would need to conquer in order to eventually capture the city. From its heights, heavy Mexican artillery could rain shells down on any American attempt to move up the narrow causeways leading to the city gates on that side of the capitol. Down in front of Chapultepec, closer to the Americans, was a set of buildings including a mill built among its arches, known locally as the *Molino del Rey* or "King's Mill." The complex of buildings was arrayed in a long-curved line, following an aqueduct. He was informed that the Mexicans had constructed facilities there for casting new Mexican cannon, and that they were melting down the church bells of the city, and therefore time was of the essence if the process was already underway. His plan was to storm the position at dawn, destroy the buildings and fortifications and then fall back before the large caliber guns of Chapultepec could focus on the American forces below them in broad daylight. He chose General Worth's division as the main force, with some support from other divisions, if required.

The role of the cavalry units not out guarding the southern and eastern flanks of the army would be to protect the western or left flank of General Worth's infantry from Mexican cavalry as the attack went in that morning. Without American cavalry screening their left flank, the Mexican cavalry might try to get around and behind the Americans. There was plenty of Mexican cavalry out there. Mexican General Juan Alverez had approximately three to four thousand cavalrymen out to the west of the Molino, and they could not be allowed to close in. Only two things would stop this from happening. A large *arroyo* (ravine) that was long enough and deep enough to keep the Mexican lancers from immediately sweeping over the Americans, and the willpower of the combined forces of all Scott's cavalrymen able to sit in a saddle that day as one force. The Michigan men of Company K who could still mount a horse were included, and the names of some of those who were there that day are known.

As General Worth's assault columns went into battle at 5:45AM to dislodge the Mexican forces from the "King's Mill," they might have noticed the formation of dragoons off to their left. The formation was small in comparison to the thousands of lancers massing in the distance. Cargill recalled that early morning.

> About 12 o'clock at night two of our cavalry (myself being one of the number) commanded by Major Summers, (Sumner) were formed in the road at Tacubaya remaining here all night and early in the morning we marched to the field of battle and arrived there just at the dawn of day.[100]

Major Edwin Vose Sumner of the Second Dragoons commanded elements of the 1st, 2nd, and 3rd Dragoons as well as one company of the Mounted Rifle Regiment that was actually mounted, (most of the companies served as infantry) along with several aides de camp and staff officers who thus far had had no opportunity for glory. Specifically, six companies of the 2nd Dragoons, one company of the 1st Dragoons, one company of the Mounted Rifles and one company of the 3rd Dragoons, reported as 270 men total.[101] The cavalry reserve sat on the far left of General Worth's forces waiting for dawn. As it became light enough to see, the combined companies of cavalry could see they were heavily outnumbered as they looked across the arroyo on the far-left flank of the attack, each company would only have had approximately thirty men, down significantly from the one hundred they were supposed to have.[102] There were thousands of Mexican horsemen, regular cavalry regiments and many units of lancers, with their different colored pennants waving from the tops of their lances. Cargill continued,

> And now our cavalry were ordered to charge upon them, a force twenty times that of our own. As

---

[100] Cargill, December 22, 1847.

[101] Hamersley, 20.

[102] Johnson, 207. The author writes that it was "Sumner's Second Dragoons," but it begs the question how Company K suffered combat casualties on that day if they were not there as well.

we approached them, passing closely to, and directly in front of the enemies batteries, grape and musket shot flying thick and fast, our men and horses falling wounded and dead, and expecting to have a severe contest with them, at this moment they were seized with fright, their columns were broken, and they fled before us in confusion and dismay.[103]

It was not so easy as portrayed by the young cavalryman. Unfortunately for the Americans, in order to get to the other side of the arroyo and come in contact with the enemy, they would need to move forward within musket range of the Mexicans and cross a bridge which lay directly in front of the Casa Mata, a fortified building full of Mexican infantry overlooking it. The Casa Mata was the anchor of the right flank of the Mexican defenses at the Molino and was also perched on the edge of the arroyo. It was a terrible bottleneck, and the horsemen were forced to ride directly at the heavily defended Mexican position while being shot at. The column then swerved left to cross over the bridge, packed together as they crossed as quickly as they could. Looking down from the Casa Mata, the Mexican soldiers could not miss. Once the Americans got over the bridge, as they faced the Mexican cavalry in front of them, the Mexican infantry in the Casa Mata continued to fire at them from behind. It was a terrible position. Once they were past the bridge, they continued forward toward the Mexican cavalry merely to escape the fire of the Casa Mata coming from behind. They were now face to face with masses of lancers directly in front of them.

The Mexican cavalry under General Alverez could and should have simply swarmed over them and wiped them out at that moment but they did not. The opposite happened, and the Mexicans fell back as the Americans moved forward. What caused the Mexican cavalry to break, and run may never be known for certain, Cargill and the other dragoons believed it was their fearlessness and of course took full credit. However, it may have had something to do with one of the American artillery batteries unlimbering at the edge of the arroyo and firing across it, tearing great holes in the ranks of the Mexican

---

[103] Cargill, December 22, 1847.

64

horsemen from their right-hand side. Regardless, what the American

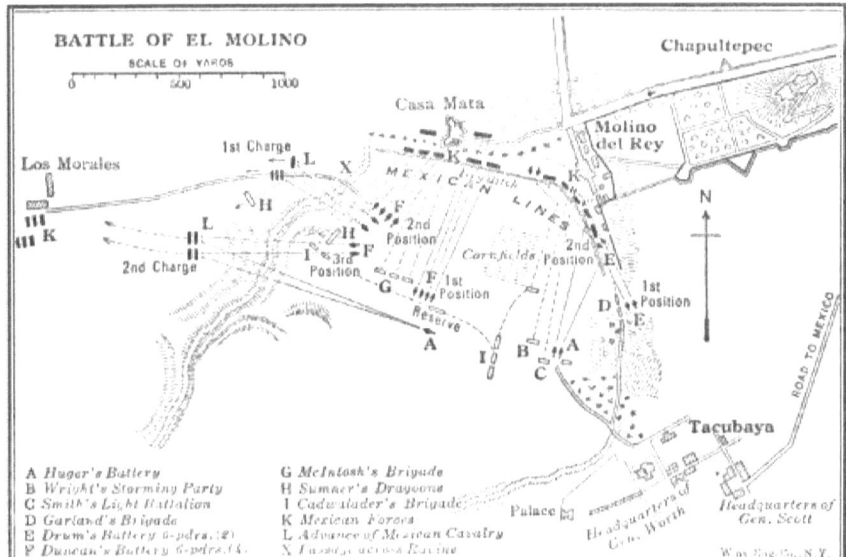

Molino del Rey. The letter H reflects the movement of Sumners Dragoons, and the letter X shows the location of the bridge.

(The War with Mexico, Vol. II, Justin H. Smith, 1919.)

cavalry did that day was incredibly brave, and they went forward despite the odds.

After an incredible fight, which went on far longer and involved all of the reserves that General Scott had assigned that day just in case they were needed, the Molino del Rey fell to the Americans, and so did the Casa Mata. The Mexican forces defending the mill fought bravely but once the retreat started it became a rout. There were four Mexican cannon that were captured, and the Casa Mata exploded in a huge fireball of captured Mexican Ammunition after the American cavalry had safely recrossed the bridge and returned to friendly lines. It was the costliest battle the Americans fought at Mexico City, and many wondered if it was worth the price paid in American lives. Colonel Sumner reported having lost seven soldiers killed, five officers and thirty-three enlisted men wounded.[104] Twenty-seven horses were

---

[104] NARA Colonel Sumner's Report to General Scott, September 20,

killed and seventy-seven wounded. During this fight Private Charles Diltz and Second Lieutenant J.C.D. Williams of Company K were wounded. The number of Company K men left capable of serving after that battle are unknown.[105]

The Battle of Molino del Rey may be seen as the final maximum effort of the cavalry elements in General Scott's Army as a whole during the campaign. By the end of the day on September 8, 1847, almost twenty percent of the total force were on the casualty list, as were the horses. Even though the losses were spread across the different companies, this effectively took them out of any offensive action at least until the slightly wounded could recover. The next phase of the battle would not require their efforts. The conquest of Mexico City itself would be carried out by the infantry, artillery and engineers.

From September 9 to September 14, 1847, the different elements of the cavalry sorted themselves out and watched the bombardment and assault on Mexico City from a distance, without playing a major role in it. The 2nd Dragoon Regiment was posted near its old position on the near side of the ravine near the bridge at Molino del Rey, again watching the left flank of the army and keeping roving bands of Mexican lancers at bay. On September 11 heavy artillery positions were dug and all day on September 12 different positions near Chapultepec were bombarded. The citizens of Mexico City, who were safely out of range, sat on the rooftops and city walls and watched the spectacle. On September 13, 1847, the assault on Chapultepec and the fight for the city gates began at dawn. By late morning the capture of Chapultepec Castle was complete, and the 2nd Dragoons were ordered to follow General Worth's troops up the causeway toward the San Cosme Gate for a time, however congestion on the single road caused General Scott to redeploy them in an arc to the left and rear of the Army's bases at Tacubaya and keep the rear areas safe.[106] As Captains Kearny and McReynolds, who were now listening to the great battle

---

1847.

[105] In his thesis, Smith wrote that Privates Edward Curtis, Augustus Dessel and George Duver were killed, and Private Cowden was wounded. This may be so, but they are not found on the rolls of Company K., 35.

[106] NARA Colonel Sumner's Report to General Scott, Sept. 20, 1847.

66

from their hospital beds had learned, capturing city gates and buildings was not a job for the cavalry. Private Cargill described watching it all but did not state that he or Company K or General Scott's Bodyguard combined, had any role in it until the next morning. Other elements of the cavalry were posted out on the right flank to screen the army as well. The battle for the city gates went on all day and into the night.

At dawn on September 14, 1847, the smoke cleared, and the Americans found themselves inside the gates of the city in two different locations. The city was relatively quiet, and it seemed that Santa Anna was either lying in wait to fight street by street, or he had evacuated the capitol. Within an hour it was clear that the latter was the case, the Mexican army had, in fact, evacuated the city, and at the same time opened the city jail to allow its inmates to cause maximum chaos. All of the cavalry that was available escorted General Scott into the city and paraded on the grand plaza in the center of town that morning. The army marched in from two directions and congregated around the main buildings in the plaza. In the famous painting of Scott entering Mexico City, it is worth noting that some of the cavalry formed behind him on the plaza would have been made up of the grey horses of Company K.

Scott enters the plaza at Mexico City September 14, 1847.
(Copy of a print-Library of Congress)

67

Colonel Sumner and the 2nd Dragoons rode in with Scott as well, Sumner mentioned the men staying up all night cleaning their uniforms and polishing their weapons. When chaos broke out because Mexican guerillas started firing at soldiers from rooftops and windows later in the day, the 2nd Dragoons were ordered to clear the plaza of people, which they did.[107] The cavalry were then billeted as residents of the National Palace, and Cargill wrote that he thought he was very lucky to have the quarters he did.[108]

Lieutenant Moseley recalled that after the fall of the city, the army began to take over and move into the major buildings on the plaza.

> On our entrance, our regiment, (the 3rd Dragoons,) took quarters in the 'National Palace,' on the 'Grand Plaza,' and literally, actually and physically reveled in the 'Halls of the Montezumas.' It was like entering an enchanted palace, such as one reads of in the Arabian nights.[109]

Company K spent the next few days hunting down snipers and keeping the peace on the streets of the city. It took nearly a week for things to settle down. For weeks afterward, soldiers were told not to travel alone or be out on the street alone at night, or they would be found with their throats cut.

Holding Mexico City did not mean the war was over, it did not even mean the route from Vera Cruz was any safer. Things were still very dangerous, and the Mexicans had not surrendered as some thought they would upon losing their capitol city. General Santa Anna and his army had left Mexico City during the night but since they weren't surrendering, where had they gone? Santa Anna had not simply disappeared into thin air. He and his army had moved to a new target, the city of Puebla, Scott's only connection to his lifeline back on the coast. Within days the Mexican forces had infiltrated parts of the city and after cutting off all routes of escape, called upon the

---

[107]NARA Colonel Sumner's Report to General Scott, Sept. 20, 1847.

[108] Cargill, December 22, 1847.

[109] Moseley, 12.

68

Americans to surrender. Colonel Childs was not about to turn over the hundreds of sick and wounded American soldiers to the Mexicans. It was common knowledge that General Santa Anna had executed American prisoners of war before, most notably at the siege of the Alamo during the fight for Texan independence, and so Childs logically refused. His exact words are unknown but later correspondence mentions that he said something to the effect that "he must decline, stating as a reason, that Americans were not inclined to do such things."[110] For several weeks Santa Anna attempted to crush the forces there by attacking one and then the other heavily defended strategic blocks controlled by the Americans, who had armed every sick or wounded patient that could hold a musket. Trapped in the city along with the garrison was Captain Ford and Company D of the 3rd Dragoons.[111] D. H. Hill stated that he had read in a newspaper account of the official report that Captain Ford had only fifty dragoons in his company at the time.[112] Although many of the men remain anonymous, it is known that one soldier of Captain Ford's Company D was wounded during the siege.

General Scott did not march out with part of his badly damaged army from Mexico City to raise the siege of Puebla, correspondence at the time suggested the rain and bad roads precluded any major movement back across the mountains to rescue the city by any large force he could put together. More likely it was simply the fact that he could not hold Mexico City and take a large chunk of his rapidly dwindling forces out of the city to go and fight somewhere else.

General Santa Anna may have wondered why the Americans at Mexico City weren't coming back to Puebla to save the garrison. General Scott was counting on someone else to solve his problem and lift the siege. Weeks before, on August 6, Major Folliett T. Lally of the 9th U.S. Infantry Regiment, had departed Vera Cruz with a large force of new troops and replacements destined for Mexico City (this included the Michigan boys of company G of the 15th U.S. Infantry Regiment.) The convoy ran into trouble with Mexican ambushes most

---

[110] Albert G. Bracket *General Lane's Brigade in Southern Mexico*. H. W. Derby & Co., Cincinnati, Ohio, 1854. 114.

[111] Albert G. Bracket, General Lane, 112.

[112] Hughes, 160.

of the way to Perote, after being stopped for two days at the National Bridge and they were forced to fight a significant battle there. By the time Lally arrived at Jalapa on August 20, 1847, the same day as the fighting at Churubusco, they had suffered ninety-two casualties.

They remained at Jalapa until reinforcements arrived back down on the coast, in the form of General Joseph Lane. He was one of the "Volunteer Generals," that had been appointed by the President. Lane had 1700 men at Vera Cruz and 150 wagons. He began moving toward Puebla with a major wagon train on September 19, 1847, unaware that it had been besieged until a few days later.

Arriving at Jalapa, General Lane took control of Lally's column and both convoys merged into a larger force containing infantry, cavalry and artillery, numbering approximately three thousand reinforcements. Arriving at Perote, he added Colonel Wynkoop's troops to his order of battle, and after several days of marching, but little interference, they met part of Santa Anna's army at the city of Huamantla. They fought a significant battle there and routed the Mexican forces, but not before losing one of the best cavalry commanders the Americans had, Samuel H. Walker of the Mounted Rifles. At this point Santa Anna knew that he could not keep Puebla under siege for much longer.

Meanwhile, Colonel Childs continued to fight off Mexican attempts to overwhelm the defenders. There were several Michigan men that had been unlucky enough to have been trapped in the city and had been convalescing in hospitals, Private Cargill's friend Charles H. Tower was one of them. He was a twenty-nine-year-old store clerk who had enlisted in Detroit and had been left behind in the hospital at Puebla on August 7, 1847.[113] For twenty-eight days every sick and wounded man that could aim a rifle was required to defend the buildings in the town, including Tower.

As news of the relief force became known, Colonel Childs sent his dragoon company out through the Mexican lines to link up with General Lane and show him the best way to break through the Mexican siege lines from outside. Captain Ford had all of the men able to mount a horse prepare themselves and they made a run for it. They ran a gauntlet of fire from the Mexicans and relied on speed to carry

---

[113] There is no further record. He may have died there.

70

them quickly past the dangerous areas. In a letter to Captain Ford written almost three years later by General Lane, he recalled how,

> You came out with your company under a heavy fire from the enemy, in fine order, and met me two miles from the city. You made known to me that the enemy were in strong force and would attack my column near the bridge, which is near the edge of the city, and that in a street fight you could render little service. I found it just as you told me.[114]

The Siege of Puebla had lasted twenty-eight days. On October 12, 1847, the Mexican forces melted away, and General Santa Anna went off into the mountains to regroup and rebuild his army. General Lane's men went from street to street and house to house to make sure the city was clear and those sick and wounded men who had been required to fight were sent back to the hospital. This event also reopened the lines of communication between Mexico City and the sea. Colonel Childs lost eighty-seven men killed during the siege. Many more were wounded, some for a second time, since they had been in the hospitals there recovering from previous wounds. Exactly how many men from the 3rd dragoons and which companies they were from (other than those likely serving with Company D.,) cannot be determined with precision.

---

[114] NARA General Lane's Letter to Captain Ford, January 14, 1848.

CHAPTER SIX

The Occupation of Mexico.

*We learned that Genl. Lane with Col. Hays Regiment of
Texas Rangers and some four companies of the 3rd Dragoons
had gone out in search of Guerrilleros. Tis a fine match,
Mustang against Guerrillero! I wish the Mustang General
every success.*
-Lieutenant D. H. Hill[115]

The survivors of the 3rd Dragoons took part in several operations chasing down remnants of General Santa Anna's army in the months that followed, while serving in a conglomerate of the four other companies. Even before the Michigan men were officially reassigned to other companies, there is evidence that they were temporarily attached. It is logical that Captain Ford would have filled his ranks with any of the Company K men who had been in hospital or convalescing in Puebla during the siege and were then available. At 11AM on October 19, 1847, some of the members of Company K may have taken part in General Lane's expedition in pursuit of Joachim Rea and Padre Celedonio Dominico de Juarata, two of Santa Anna's ablest cavalry commanders. Lane's force included five infantry units, two artillery batteries and Captain Ford's Company D of the 3rd Dragoons. According to General Lane,

> The advanced picket of the enemy lancers was discovered about four o'clock PM, and, after a halt to allow our cavalry to come up, the advance was

---

[115] Hughes, 168.

72

resumed, and in a very few minutes our cavalry was 'ordered to charge,' when a running fight was kept up for a mile and a half, after which the lancers halted on a hill in a strong position and made a sharp fight until our artillery and infantry arrived. Then they took flight. The dragoons followed and kept the enemy engaged for four miles farther and until within a mile and a half of Atlexco. Here the whole body of the enemy was found posted in a strong position, protected by chaparral, hedges, etc., but our cavalry dashed among them, cutting them down in great numbers.[116]

One Private from the 3rd Dragoons was killed and one wounded in those operations, but those men were not from Company K and therefore were not from Michigan.

It was during those weeks after the major fighting around the capitol city and at Puebla had ended that many members of the regiment became sick. The hospitals in Mexico City were set up in major public buildings such as churches and monasteries. Each division of the army had its own suburb it was based in and each had its own hospital. As if to emphasize the power of nature over the sick in their beds, many in Mexico City also mentioned that an earthquake occurred while they were there.

During the months of September and October, the following Michigan men were discharged from the army at hospitals in the vicinity of Mexico City or died there:

Thomas Bruette, a twenty-six-year-old Private who was a tinsmith from Canada, died of disease at the hospital in Mixcoac on September 12, 1847.

Chandler Ceady (or Kailey) was an eighteen-year-old Private from Tecumseh who died in Mexico City on September 23, 1847.

Avery Ellis, a twenty-five-year-old Private who was a blacksmith from Tecumseh, died in Mexico City on September 24, 1847.

Private Orville Royce of Tecumseh, died of disease in Mexico City on October 9, 1847, aged twenty-six.

---

[116] Bracket, 25.

Private Allen T. (or F.) Welch, age twenty-two, was discharged for disability at Mexico City on October 26, 1847.

Julius A Holmes, First Sergeant of the Company, a twenty-four-year-old soldier from Tecumseh, was discharged for disability in Mexico City on October 27, 1847. [117]

Private William H. Patten was discharged for disability at Mexico City on October 30, 1847

Samuel P. Cargill wrote to his brother from Mexico City in Detroit on October 27th, 1847, that,

> There have been no mails back to Vera Cruz in a very long time. The only way that General Scott has had of sending dispatches is by sending Mexicans, (hired for the purpose) who traveled not on the main road, but across the mountains and byroads to get to the place to which they were sent.[118]

None of the newspaper men from Company K were sending letters back to be published, so how did Cargill get a letter out? Were any of the self-proclaimed "Printers" left to write a letter home? The group of gentlemen who had committed to writing back and publishing the exploits of the 3rd Dragoons from the "Theater of War" had nearly disappeared by the time Mexico City had fallen. They did not seem to be sending anything back. A.T. Welch and William Patten were so incapacitated by fever and dysentery that they were both discharged for disease and sent home as soon as a convoy could make it safely back to Vera Cruz in November of 1847.

When Welch and Patten left, their friend and fellow printer Daniel Cruice lay deathly ill in the hospital, too sick to move. It appears that only Charles Burnham made it through the entire ordeal and there is little record of information sent by him.

First Lieutenant Moseley wrote that "On the 1st of November 1847,

---

[117] The December 28 Newspaper article describes him as the Orderly Sergeant.

[118] Cargill, December 22, 1847. He had become a Surgeon's assistant and so had time to write and access to the materials to do so while in Mexico City.

the first wagon-train of sick and wounded officers was sent down to Vera Cruz and from thence across the gulf home to recruit."[119] Cargill also wrote about that train, the first to head back down since the fighting of August and September had ended, "the principal object of which is to send home the sick and wounded. Among the number are Captain McReynolds and Lieutenant Brown, of our company; also, Welch, who was a printer with Bagg & Harmon." He also mentioned his friend, Private Charles Tower.

> I wish you to say to Mr. Town that Charles is now in Puebla. He had been sick for a while before the army started from that place and had not sufficiently recovered to come with us. He will doubtless be up with the next train, which is expected in a few days.[120]

It is unclear of private Tower ever did make it to Mexico City. It seems either Private Cargill was part of the convoy security and rode down to Vera Cruz himself on November 1, 1847, or he was able to send a letter back with one of the sick and wounded who were evacuated. Those wounded and sick officers who were sent home were put on recruiting duty following their convalescence in order to refill the heavy losses of the regiment. Those who could not be moved remained in the hospitals to recover or else they died of their illness.

After the convoy left, two more Michigan men who should have gone with it but were too badly wounded or sick succumbed to their injuries:

Private Daniel Axford died while in the hospital at Puebla on November 9, 1847.

Private Daniel Cruice one of the last two "Printers," died of disease in a hospital at Mexico City on November 12, 1847.

Captain McReynolds left his beloved Company K in the hands of Second Lieutenant Henry until Second Lieutenant J.C.D. Williams, who was also wounded, recovered enough to take command. Only Second Lieutenant Henry would have been available to see Captain

---

[119] Moseley, 12.

[120] Cargill, December 22, 1847.

75

McReynolds off as he was assisted into a wagon containing his baggage, wearing a sling around his wounded left arm and shoulder.[121] He was going home to recover, and if well enough might also attempt to recruit some replacements from Michigan to bring the company back up to strength. He brought several of those discharged for illness home with him. On November 24, 1847, he was in New Orleans with Lieutenant Colonel Moore[122], who had been second in command of the regiment, and led the Second Battalion during its time in Mexico, and had also been sent home.[123] The Second Battalion of the 3rd Dragoons was being run by his replacement, Major William H. Polk of Tennessee, the President's younger brother, who had just arrived for duty. It is unknown how long the journey down to Vera Cruz took, or if the column of sick and wounded was attacked as it headed eastward. It is known that the sick and wounded continued to die along the way and were buried along the road down to the coast or added to the existing cemeteries at the next nearest town where there had already been American soldiers buried. Few of those names and none of those locations are known or marked.

Once all the casualties had made it back across the Gulf of Mexico to New Orleans and were unloaded, there would likely have been more hospital time. The Michigan veterans then would have boarded a steamboat to take them upriver to Cincinnati, and then north to Toledo and then home. By this time there were also new rail lines between some of the cities mentioned, and if soldiers could afford it, they took advantage of this new "high-speed" technology.

Back in the United States, new technology was also passing information faster than before. On December 2, 1847, an Ypsilanti telegraph operator, O.B. Holman, sent the following information to the editor of the Detroit Democratic Free Press, which was published as quickly as it was received.

Captain McReynolds of the 3rd Dragoons, U.S.

---

[121] At least one newspaper reference states that it was his left arm and shoulder, Scott also reflects this in his report.

[122] Lieutenant Colonel Moore was a Virginian who was appointed to his position from Kentucky.

[123] *Jackson Patriot*, December 7, 1847.

Army arrived in this place last evening from the city of Mexico by way of Vera Cruz, New Orleans and from thence across the country to this place by way of Toledo.[124]

He had arrived in Ypsilanti "on the cars," as they called it back then. McReynolds had stopped there that evening (December 1, 1847,) around 7pm and expected to arrive by train in Detroit the evening of December 2, 1847. The telegram went on to say that he had left command of Company K to Second Lieutenant Francis Henry of Wisconsin[125], and that Second Lieutenant J. C. D. Williams, who was wounded at Molino Del Rey, was scheduled to take command on December 1, 1847, if he had recovered sufficiently. Williams was promoted to First Lieutenant on January 1, 1848. First Lieutenant Brown was still sick and currently at Vera Cruz in the company of several other recovering Michigan Officers awaiting transport back to New Orleans aboard the steamer *Galveston*. One of the reporters who met with Captain McReynolds in Adrian, Michigan, went into graphic detail.

This gallant officer-Captain of the Company of Dragoons from this state-passed through our village on Wednesday last, on his way home to Detroit. During a stay of an hour or two with us, he was called upon by many of our citizens, and a salute was fired on his departure. Captain McReynolds had with him the sword which was presented to him by the citizens of Detroit when he left for the war. It had seen service. The scabbard is bruised, the sword is in places 'All rusted o'er with red,' and the belt and hangings have dried blood upon them. Some of the fastenings are cut in two by grape shot.[126]

---

[124] *Detroit Democratic Free Press*, December 10, 1847.

[125] Born in Illinois and commissioned from Wisconsin, he was the only officer not from Michigan in the company.

[126] *Detroit Democratic Free Press*, December 20, 1847.

One of those who accompanied the captain home was Private A.T. Welch, known as "Tate," to his friends. The gaunt and debilitated "printer" must have seemed almost unrecognizable to his fellow employees at the Detroit Democratic Free Press, who had not seen him since the previous April. The paper described his homecoming, stating that-

> He was with the dragoons from their landing at Vera Cruz-during their tedious marches and gallant charges until they 'reveled in the halls' where, after a residence of six weeks, he was honorably discharged from the service, on account of sickness, and returned to this place. Of the conduct of the unflinching Wolverine boys, he speaks in glowing terms, and the mournful tale of havoc by death from disease and sickness, he relates, will cause many a broken heart-many a sorrowing household and many a tear of sympathy for the friends of the departed brave, and noble spirits that have fled.[127]

The people back in Detroit would have received no word regarding casualties within the company until these men returned in December. It would have certainly been a sad Christmas in both Detroit and Tecumseh for the families who learned from this small group of survivors how many had been lost, and how many more were in peril of wounds or sickness when the captain and the other veterans had departed Mexico. An article was published in the Whig Standard on December 21, 1847, in Hillsdale. This was borrowed from the Michigan Expositor.

> We do not remember the number of men who left for Mexico in this Company, April 26th, 1847, but suppose there were from 80 to 100. Of that number, we are informed less than 30 are now living: the remainder having been cut down by disease or killed in battle. Of the few still alive, 17 were able to do duty when Captain McReynolds left. This is a picture of

---

127 *Detroit Democratic Free Press*, December 9, 1847.

war.

The following have returned and are on their way home-discharged: Orderly Sergeant Julius Holmes, Privates-A. Warriner, Alden Ball, A. T. Welch, ____ Barlow, Wm. Batten,[128] Stephen Armstrong, Wm. Ayers.

Corporal John W. Leake, known to many of our readers, died at Puebla on the 9th of August, after a short illness of dysentery. Captain McReynolds when here, spoke of his death with feeling.[129]

The conquering hero was welcomed home, and he probably felt very lucky to have made it home at all, with all the death he had witnessed. It was Christmastime, and it would have been hard not to feel guilty when he thought of the survivors of the company still in Mexico, far from home. Upon arrival back in the Detroit area, Captain McReynolds was invited to dinner by many groups, including a group calling themselves "The Citizens of Bloomfield." He wrote back that he took great pleasure in accepting these invitations, "for the manifestation of patriotism and kindness."[130]

One officer and sixteen men in the saddle. A sad fraction of the company. Unfortunately, all of the companies in the Second Battalion were in similar shape. The company paper strength for Company K listed for November and December was 83, but seven were discharged for disability and five listed as having died during the period. The majority were in the hospital. There really weren't enough men in the saddle to be assigned to one of the garrison forces being discussed at headquarters.

On Christmas Day, December 25, 1847, one such garrison, a mixed detachment of troops, was sent out to occupy the town of Pachuca, near some mining interests and to protect a large number of English civilians who worked there, north of Mexico City. Two dragoon

---

[128] There was no William Batten in Company K of the 3rd Dragoons. The news reporter may have confused him with Franklin Bates, who had been released from the service and was on his way home.

[129] *Hillsdale Whig Standard*, December 21, 1847

[130] *Oakland Gazette Pontiac*, December 18, 1847.

companies, were sent, one of them Captain Gaither's Company C of the 3rd Dragoons. Four companies of the 9th Infantry under Captain Withers, and one battery of artillery under a certain angry regular army lieutenant, one Daniel Harvey Hill also went. Except for his artillery section, all of the other units in the detachment came from the "Ten-Regiment Bill." In particular, the 9th Infantry Regiment was nicknamed "The New England Regiment" because all of the companies had been raised from volunteers in the New England states. Just a few months earlier they would have all been referred to by Lieutenant Hill as those "vile volunteers." Even though they had withstood shot and shell just as he had, in Hill's mind they were still "vile". Clearly, he was going to have to learn to keep his thoughts to himself and get along if he was to survive at this outpost surrounded by those whom he felt were less qualified than he was. He was not happy about being posted sixty miles from Mexico City, never mind his friends in the regiment. Having to be collocated with these people was more than he could stand. He stated in his journal, "My disgust at finding myself permanently separated from my regiment is infinite."[131] One week later he was joined by eleven more cavalrymen as Michigan men transferred from Company K arrived to reinforce Company C.

---

[131] Hughes, 159.

80

CHAPTER SEVEN

Redeployment, Replacements, and Discharge.

As one of the Kentucky Captains says, the General &
Colonels are going home to run for office and so they let the
boys do as they choose.
-Orlando B. Wilcox[132]

In January of 1848, General Scott consolidated the army that had conquered Mexico City. Many volunteer units were arriving, and small groups of replacements were arriving from recruiting stations back in the United States to rebuild the badly depleted "regular army" regiments. The regular cavalry, that which was left of them from the 1st, 2nd and 3rd Dragoons, were still in Mexico City at the Palace. Kearny's Company of the 1st Dragoons remained General Scott's bodyguard, but Company K from Michigan was dissolved, and the survivors were transferred to fill the other companies of the 3rd Dragoons in the Second Battalion, C, D, E, and G. There were too few of them who were fit for duty and only one officer left in Company K, but Second Lieutenant Henry may have been sick as well. The 2nd Dragoons also consolidated their companies. No replacements were coming for the dragoon regiments any time soon. This explains why records from Company K are hard to find.

Most of the men went to Company D. The Indiana company gained eighteen from Michigan. The second highest number went to

---

132 Robert G. Scott, Forgotten Valor, *The Memoirs, Journals and Civil War Letters of Orlando B. Willcox*. Kent State University Press, Kent, Ohio & London, 1999. 112.

the company from Kentucky, Company C. They gained eleven Michigan men. Only a handful of soldiers went to the New York Company E, just four men. The Louisianians probably did not know what to make of the three Michigan men who joined Company G.

Being assigned to a different company did not stop the soldiers from getting sick and being sent home.

Private Edward W. McIntosh was discharged for disability from the hospital at Vera Cruz on January 9, 1848. What became of him after that is unknown.

Private Joel Parish was transferred to C Company along with his younger brother Enos on January 4th, 1848, while at Mexico City. Ten days later, on January 14, 1848, Joel was discharged for disability and sent home.

As previously stated, when the reorganization occurred at the New Year, Captain Ford of Company D received twenty-two permanent transfers, eighteen from Company K., bringing his strength up to 81 men in the saddle. His unit was now one of the strongest. Wherever Company D went from January to June, twenty-five percent of the unit was made up of Michigan veterans.

As mentioned previously, Major William H. Polk had arrived as the "second" or "extra" major authorized for the regiment, and from that point on he led elements of the Second Battalion of the 3rd whenever they were sent out on operations from that point on. Captain Ford, the captain of Company D., was breveted to Major and worked on General Lane's staff coordinating cavalry support for anti-guerilla operations for a few weeks and then was sent home on recruiting duty to Indiana.

On January 18, 1848, General Lane headed out of Mexico City in search of Santa Anna himself, and among the units in his column, were two companies of the 3rd Dragoons commanded by Major Polk. These were very likely an amalgamation of whoever was able to be present for duty from the four remaining companies. They attempted to capture the Mexican General at *Tehuacan* but missed his departure by approximately one hour, and their horses were too exhausted to go on. Although they missed General Santa Anna, they did capture a large amount of his family's personal baggage.[133] Two of the Tecumseh

---

[133] Bracket, 236

boys, James Simonds and Carlisle Soper claimed to have also been on that patrol and appropriated a boot of the Mexican General, which they apparently carried home as a souvenir of the occasion.[134] During the return trip, Lane recalled,

> Permission was given to Major Polk to take two of his companies and recapture some United States animals reported to be some two leagues away. Major Polk returned in due time with one American horse and twenty-six mules, belonging to the United States.[135]

In February General Scott was replaced by General William O. Butler, and he oversaw the peace negotiations while continuing the policy of occupation of major cities and strategic locations along the route from Mexico City to Vera Cruz, including opening a second route that led in a southern arc through the cities of Orizaba and Cordova. Now there were two routes to cover, and guerilla forces operated along both.

On February 5, 1848, and again on February 17, 1848, Lane's force again went out to fight the guerrillas and keep the roadways safe. Major Polk commanded the "company" of the 3rd again, but also one of the Mounted Rifles. It is unknown which company went but Lieutenant Adde of Company G led the men on those patrols.

The last expedition conducted by General Lane before the armistice went into effect on March 5, 1848, also included Michigan soldiers of the 3rd Dragoons although it is unknown how many. By the names of the officers mentioned leading the units, it appears they were Company E led by Lieutenant Diwer, and Lieutenant Adde of Company G, however they were filled with so many men from Company K they were recognized as such at least once in official reports, although they technically no longer existed. They were attempting to capture Padre Juarata at *Zacualtipán*. They charged into the town with a group of Texas Rangers in the lead. Lane states,

---

[134] Bonner, 643.

[135] Bracket, 29.

A running fight ensued, in which the Padre lost about thirty men killed. Hays, after this, returned to the plaza; there met Second Lieutenant Alexander Hayes, Eighth Infantry, acting assistant adjutant general to General Lane, who informed him that the general was, with a small force, hotly engaged with superior numbers, and Colonel Hays ordered Captain Daggett's company of rangers to his assistance. The Mexicans fought bravely, taking advantage of their position and firing under cover of every object that gave protection, but were overcome with a loss of thirty killed. While Hays was in pursuit of scattered fugitives, he met Major Polk with two companies of the 3rd U.S. Dragoons, when they arrived in the suburbs of the town. Hearing firing, the Major dismounted Walker's Rifles and directing them against a cuartel from which the firing was coming. They silenced it, then moved on to a nearby church from which fire occasionally came. Lieut. George E. Maney, the Third Dragoons, Acting Adjutant, was sent back to order up the two companies of dragoons, with which Polk advanced beyond the plaza, to the top of a hill to the left, which dispersed the enemy. Polk ordered Lieutenant Adde [Company G.] to remain in occupation of the plaza, and Lieutenant Diwer's company [Company E.], divided up into small squads, to pursue the enemy. Meanwhile, Walker had taken the church, killed some of the enemy, he losing only one horse.[136]

After the armistice was signed in March, there was not much more in the way of patrolling, the men simply wanted to get things wrapped up and get out of there. The companies of the 3rd Dragoons were primarily on convoy escort duty, between Mexico City and Vera Cruz, as the sick were sent home, and supply depots cleared out. As much as Lieutenant D. H. Hill of the artillery disliked the "volunteers," perhaps

---

[136] Bracket p. 31-32.

84

he did not notice that the cavalry that safely escorted his wagon train to Vera Cruz so he could go home on sick leave was none other than Company G, 3d Dragoons, under Captain Duperu and Lieutenant Moseley, with three of the Michigan boys riding along beside him.[137] Even as their departure grew near, the men of Company K continued to dwindle away, through combat losses or illness, although their names were now reflected on someone else's orderly reports.

It appears that by the time the war ended, each of the companies was blended together and led by one of the two remaining captains, Duperu or Gaithers, depending on who was able. On any particular mission, they were most likely led by one of the surviving second lieutenants. It is not unrealistic to say that their operational strength at the end of the war would have been one company instead of five. One hundred men out of five hundred.

For Brevet Major McReynolds, who had more or less recovered, and had been breveted for his bravery at Churubusco, the news would have been very disturbing. He had not been required to resign, although he had lost the use of his arm to a great extent and desperately wanted to get back to his regiment as the new year began. He had thought First Lieutenant Williams was commanding the company but discovered at some point that five days after assuming command, on December 6, 1847, orders arrived for the wounded officer to depart for the United States and recruiting duty.[138] He now had no company to command, and his remaining officer, Second Lieutenant Henry, who may also have been sick, and was not from Michigan, would have been assigned elsewhere. It is probably for this reason that he began a recruiting drive in Michigan for several months in an attempt to justify the recreation of his beloved Company K. It was not to be. Only the swift return of Major McReynolds, with a large number of recruits might have resurrected it, and that did not happen in time.

In May, two companies of the 3rd under Major Polk's command escorted the delegation to ratify the peace treaty, ending the war, so some Michigan men may have been on hand to have witnessed the event. On June 2, 1848, Orlando B. Wilcox, a "Regular Army" officer

---

[137] Hughes, 173

[138] NARA Letters Received from the Adjutant General April 10, 1848.

from Detroit, serving with Company G of the 4th U. S. Artillery Regiment, accompanied part of the army down to Vera Cruz, providing artillery support to the two companies of dragoons escorting the train. Although he had himself graduated from West Point only ten months earlier, he also had nothing good to say about "volunteers." He wrote in his journal,

> Encamped near Rio Frio. Last night rain broke up the volunteer's camp. Some got up in their shirts & drove tent pegs; some got up & cursed & laid down; & others turned out by comp[anies] (3rd Dragoons) & danced round a huge fire. To-day three men were killed & stripped by the bridge between Buena Vista and Cordova 2 (Kentuckians) and 1 of our teamsters.[139]

Perhaps Wilcox did not realize that the 3rd Dragoons had been through all of the major battles around Mexico City and seen far more combat than he. Perhaps he might have been more respectful, even if he did consider them volunteers. Perhaps he knew that Company K was from his hometown. He did pal around with some of the officers of the 3rd, and he mourned their loss when death came suddenly. On June 13, just a few days later, as they were encamped at Paso de Obejas, two days from Vera Cruz, he wrote in his journal. "Found Captain Alphonse Duperu's Company of the 3rd dragoons at the National Bridge. Martin, with whom I frolicked with but yesterday at Jalapa, is dead.[140] Second Lieutenant Martin had been the only Indiana officer physically fit to take Company D out on patrols. First Lieutenant Schoonover, who appears to have been ill most of the time, now led the company where most of the Michigan men were serving. They made their way down to Camp Washington and waited for the ships to take them home, on the beach near Vera Cruz.

The Second Battalion, 3rd Dragoons officially departed Mexico City on June 12, 1848. Lieutenant Moseley of Company G was there and described how the night before there had been a huge firework display

---

[139] Scott, *Forgotten Valor*, 108.

[140] Ibid, 115. Lieutenant John W. Martin. Company D, 3rd Dragoons.

in the central plaza, and it had ended with the illumination of giant letters in front of the palace spelling out "PEACE." He was still attached to Worth's Division, and they were the last ones out of the city. He also recalled the final ceremony that morning, as soldiers from both armies were arranged on the plaza.

> The drums roll a salute along the line; the guards present arms; the cannons flash a salvo; down flutters the Star-Spangled Banner from the flagstaff on the National Palace; up waves the red, white and green drapeau, the tri-color of Mexico; the transfer of sovereignty is made.[141]

On his march down to the sea, between Puebla and Perote, Lieutenant Moseley witnessed an incredible occurrence in his convoy, proving that once again, nature hadn't given up on attacking the Americans, even if the Mexicans had.

> A thunderstorm passed over the column of troops. The electric currents were strong, the muskets first-rate conductors. The consequence was a stunning report, and a whole company of infantry was stricken to the ground, stunned, paralyzed, and blinded, some lacerated with wounds, others burnt and scorched; happily, none killed.[142]

The 3rd Dragoons left in pieces, different companies sailing from Vera Cruz with whichever division they had been assigned to. Which officers were leading which company and when is also questionable, likely those who could maintain a saddle. The scant records for the 3rd Dragoons state that they officially left Mexico on July 2nd, 1848, although this may reflect the northern battalion under General Taylor and not those who had fought for General Scott.[143] Only Moseley left

---

[141] Moseley, 17.

[142] Moseley, 17.

[143] NARA Letters Received by the Adjutant General, Army Returns, July-August, 11.

a reflection of what would have been a common memory for each of the cavalrymen of Company K., once they reached Vera Cruz.

> We march through the old stone-girt city, embark on the different transports there waiting our arrival, and leave the low sandy shore far behind.
> High up in the heaven looms the ever-snowy peak of Orizaba, the setting sun gilding his rest with a sheen of dazzling glory.[144]

The remains of the 3rd Dragoons most likely disembarked at New Orleans. Whether or not they joined the northern companies of the 3rd Dragoons there is unknown. The survivors were transported upriver on steamboats and disembarked at Jefferson Barracks, Missouri, in pieces.

Meanwhile, recently promoted Brevet Major McReynolds was frantically attempting to rebuild his non-existent company. In April and May of 1848, he had been able to gather recruits in Michigan to replace the losses of Company K during the previous year. The following advertisement was printed in a Kalamazoo newspaper:

> Captain McReynolds, U. S. Dragoons, is now recruiting in Detroit, for dragoon service, in the U. S. Army," and that "A good opportunity is thus presented to our young Wolverines who may desire a bold and dashing service, good pay and a farm of land.
> -Captain Andrew T. McReynolds, Kalamazoo Gazette, April 21, 1848.[145]

Brevet Major McReynolds and his replacements gathered in Detroit, were sworn in and possibly uniformed, and then traveled south from Michigan for a rendezvous that now had no point or purpose. These recruits account for twenty-seven of the one hundred forty men claiming to have belonged to the unit. They served for no more than two months generally and met the 3rd Dragoons as the

---

[144] Moseley, 18.

[145] Andrew T. McReynolds, *Kalamazoo Gazette* April 21, 1848.

remains of the regiment returned to Jefferson Barracks from Mexico in June and July of 1848. The twenty-seven recruits that were gathered by Brevet Major McReynolds were discharged from the army at the Barracks on July 16, 1848, having never set foot in Mexico. McReynolds had been waiting there with them attempting to equip them as the unit arrived back from Mexico. Keeping another dragoon regiment was not something the politicians in Washington D.C. were interested in, and so Andrew T. McReynolds was informed at some point that it was too late to resurrect Company K. It must have been an interesting reunion. Certainly, there were many he looked for who were not among the survivors when they arrived, and many went straight to the hospital.

Two weeks later, on July 31, 1848, the survivors of Company K were called to form up in ranks one final time at Jefferson Barracks.[146] They would have already turned in their horses and saddles, weapons and equipment. The new men, if there were still any around from the other companies, hadn't been issued all of their gear and probably had nothing but their clothing. It is difficult to know if the new recruits looked upon the gaunt weatherbeaten veterans of Company K with reverence or pity. And how did the grizzled veterans in their tattered and patched up uniforms react to the replacements, standing there in their new and spotless cavalry blue jackets and caps? Only those that were there knew for sure. The following surviving combat veterans were discharged during the last week of July and the first week of August 1848 from their transferred companies:

Sergeant Edward H. Benedict, a twenty-six-year-old from Trumbell City, Ohio discharged from Company E.

Private John Bletter, a twenty-eight-year-old from Detroit, discharged from Company D.

Private Charles Burnham, a twenty-two-year-old from Detroit, discharged from Company D.

Private William Carby, 28-year-old from Detroit, discharged from Company E.[147]

Private Thomas Castle, 21-year-old from Detroit, discharged from Company C.

---

[146] Smith, 37.

[147] Pension reference states he died at Sacramento, June 16, 1884.

Private Richard Conklin, a 21-year-old from Tecumseh, discharged from Company C.

4th Sergeant David Davis, a 34-year-old from Detroit, discharged from Company G.

4th Corporal Erastus M. Denio, a 22-year-old from Detroit, discharged from Company G.

Private Patrick Drum, a 24-year-old from Detroit, discharged from Company D.

Private William Dudley, a 33-year-old from Detroit, discharged Company D.

Private William Eagan, a 22-year-old from Detroit, discharged from Company C.

Private William Ellis, a 23-year-old from Tecumseh, was the last of the three Ellis brothers to still be serving at the end of the war. He was discharged from Company C when he was mustered out.

Private Michael Fitzgerald, a 27-year-old from Detroit, discharged from Company D.

Private Isaac Gibson, a 21-year-old from Detroit, discharged from Company D.

Corporal Edward Grey, a 25-year-old from Detroit, discharged from Company D.

Private John Heidlauf, a 29-year-old from Detroit, discharged from Company D.

Private Ranseller H. Hinman, a 21-year-old from Detroit, discharged from Company C.

Private Charles P. Hungerford, a 25-year-old from Detroit, discharged from Company D.

Private John Kennedy, a 34-year-old from Detroit, discharged from Company C.

Private Dennis R. Lynch a 24-year-old from Detroit who served as a blacksmith. He was discharged as a member of Company G.

Private Patrick McDivitt was a 28-year-old from Detroit, discharged from Company C.

Sergeant Zeba W. Merritt was a 24-year-old from Detroit. He was discharged from Company C.

Private Stephen Mosher was a 22-year-old from Tecumseh, discharged from Company E.

Corporal Jesse Myers was a 30-year-old from Detroit, discharged from Company D.

Private Enos Parrish was a 21-year-old from Detroit, discharged from Company C.

Private Lorenzo Parsons was a 24-year-old from Detroit, discharged from Company D.

Corporal Sylvester Peltier, a 19-year-old from Detroit, discharged from Company E.

Private Benjamin Raney, a 19-year-old from Detroit, discharged from Company D.

Private Justin N. Remington, a 22-year-old from Detroit, discharged from Company C.

Musician Francis S. Reno, a 20-year-old from Detroit, was discharged from Company D.

Private Ira Reynolds, a 33-year-old from Tecumseh, discharged from Company C.

Private James N.A.S. Simonds, a 21-year-old from Tecumseh, was discharged from Company D.

Private Carlisle B. Soper, a 22-year-old from Tecumseh, discharged from Company D.

Private Milton Stoddard, a 20-year-old from Tecumseh, discharged from Company D.

Private John Tabor, a 23-year-old from Detroit, discharged from Company D.

Private John D. West, a 21-year-old from Detroit, was discharged from Company D.

Thirty-seven Michigan soldiers were discharged at Jefferson Barracks who were still more or less fit for duty and had been listed as original members of Company K. These were the survivors of the fight at the National Bridge, the Battles of Contreras and Churubusco, of Kearny's Charge and the Battle of Molino del Rey. These were the men who had survived the months of patrols and convoy duty, ambush and counter-ambush and seemingly endless guerilla warfare. Seven other soldiers who had been sent back to convalesce at Jefferson Barracks from hospitals in Puebla, Perote and Vera Cruz, were discharged with them: Joshua Beaman, Clement Beniteau, Edwin Howell, David Ousterhout, William Seymour, Charles Wagoner and William Winters, bringing the total number of discharges of the survivors that day to forty-four.

It should also be said that most of them would have qualified as having been a casualty. There were many instances of these men suffering "slight" wounds that would allow a soldier to recover after a few days or a week, and many of those veterans who returned had also spent part of their time in Mexico on the sick list. It is safe to say that no soldier that served in Mexico with the 3rd Dragoons came home as healthy as they left.

Major Cass and the First Battalion were discharged as well, having also transited back through New Orleans. It is likely that, although they certainly had less combat related deaths, those caused by illness would have been similar to losses with the other half of the unit. During the months when yellow fever was most common, no doubt large numbers of Cass's First Battalion would not have been fit for service for some time.

And so, forty-four combat veterans from Company K were discharged at Jefferson Barracks. How many of them then went back home to Michigan is unclear, but fifteen can be accounted for. Those with Civil War records thirteen years later clarify the destiny of some. Others may have headed west, as the Gold Rush of 1849 was not far off. Some remained in the service, transferring to one of the other two dragoon regiments, or perhaps the Mounted Rifles, who operated along the Oregon Trail for some time after. Burial records account for some others, but those that are identified are extremely rare. To the people of Michigan, nearly thirty more of them seemed to have simply never returned.

What of Private Cargill's friend Private Charles H. Tower? It seems the clerk from Detroit, like so many others, was not among those in the formation. He was among the missing.

EPILOGUE

The obvious question is, why do we, the people of Michigan, not know about them? There are several reasons. In the case of Company K, they enlisted in a "regular army" unit; therefore, all documentation and pension records are through the federal government and not the state of Michigan. Unfortunately, there are few lasting records of the 3rd Dragoons because it only existed for a short time, even less for Company K because it was disbanded and folded into the other companies in early January of 1848. There are few records for the 3rd Dragoons overall, especially the Second Battalion, and no accurate muster rolls exist.[148] The regiment never fought as one unit. All the key leaders from Michigan were badly wounded or disabled by illness to a point that they were evacuated home, leaving Second Lieutenant Henry of Wisconsin to manage those few still left. He had little or no power to control the destiny of the company. Also, many of the enlisted men became disabled through illness and were lost track of by the company after they were left in the many hospitals between New Orleans and Mexico City. Unlike the Civil War, few enlisted men were able to get a letter home to let anyone know of friends and relations. They came home from the war alone or in small groups, there was no welcome home or grand parade. Their homecoming was rarely mentioned in the paper, and likely was tied to having to break some very bad news to other families in their small Michigan towns about those never coming back.

Another main reason is that the war happened thirteen years before something even larger and more traumatic engulfed the country, and

---

[148] Smith, 37.

so the Mexican War seems to regularly get pushed into the background in relation to the American Civil War. An editor of a prominent Detroit newspaper wrote in 1847 of the dragoons of Company K that, "Michigan will never cease to remember their deeds."[149] -and then it did. As the Civil War began, the memory of the sacrifice made by Michigan men in Mexico was quickly replaced by the memories of the trauma and sacrifice much closer to home and by so many more Michigan boys. By 1865, the twenty-seven months of fighting and hardships in Mexico seemed long ago in comparison to what the country had just gone through over five years. Many Mexican War Veterans picked up the flag yet again and represented the state during the Civil War since they were the only locals with any military experience. It was not uncommon for a Private from the Mexican War to become a Captain when the Civil War began. These veterans of two wars put the memories of the previous one out of their minds when dealing with the sheer scale of the current carnage. Many survivors of the Mexican War went on to add to their laurels during the Civil War, and for some the shadow of death that had passed them by in Mexico caught up to them in more famous places such as Antietam, Vicksburg, and Gettysburg. Regardless, memory faded with time to the point that their duty in Mexico has been largely forgotten.

Finally, they were not remembered the same way their later more predominant brothers from the Civil War were. Not long after the end of the Civil War, veteran organizations sprang up across the north, most notably the G.A.R. The "Grand Army of the Republic" was created to lobby for pension and widow's relief and support for Union soldiers and their families. Within their ranks, the veterans of that earlier war were remembered for their service against the Confederacy, not Mexico, and were buried and memorialized as Civil War veterans. Once organizations existed in the south to remember confederate veterans, Mexican War soldiers became lost in that crowd as well.

There was a short-lived attempt at recognition. In the 1870's there was a small organization of Veterans of the War with Mexico in Michigan that held together for a few decades before it disappeared forever. The records of this organization are sparce, much of it was lost

---

149 *Detroit Democratic Free Press*, December 9, 1847.

94

in a fire around the turn of the century.[150] They organized for the same reason that Civil War Veterans were organizing in the postwar years. They felt their service and sacrifice should be rewarded with a pension, which was not unreasonable considering how many of them came home with wounds and lifelong debilitating illness. They wanted a government headstone, which all who had already passed did not have, and perhaps a flag for Decoration Day, which was becoming an informal national holiday. It was also a time to remember and be remembered. For a time, they held in mind that earlier war, and memories of a younger country, when the State of Michigan still seemed brand new, when the ideas of railroads and the telegraph were just beginning to spread across the nation; and the darker, future-horror of "The War Between the States" still lay in the future.[151] Few of the men have a government headstone that says "Mexico" on it.

It is both sad and unjust that, for most of these brave men and their families, acknowledgement of their previous service and sacrifice even now goes unrecognized and without memorial. Hence this series of books.

The second book in the series covers all of the other soldiers that enlisted from Michigan and served in the infantry and artillery of the Regular Army, to include the Voltiguers & Foot Riflemen Regiment, as well as the Mounted Rifle Regiment.

The third book in the series covers the odyssey of the *original* 1st Michigan Volunteer Infantry Regiment. Not to be confused with the Civil War regiment from fourteen years later, this was the first full military unit from the State of Michigan to leave the country and fight overseas. There were nearly one thousand Michigan men in the First Michigan Infantry Regiment, half of them became casualties, and they

---

[150] It is my understanding that the records were placed in a municipal building in the Petosky or Harbor Springs area by Captain Toll of Company E of the 15th US Infantry Regiment for safekeeping as the organization had ceased to function. The records burned with the building around 1899, and they were gone forever.

[151] Southern Veterans of the Civil War were very active in lobbying for a pension from the Mexican War, since all were ineligible for that which existed for the Union veterans. This pension did finally arrive, but only after many northern politicians argued against it for the very reason that it would go to men who had later been Confederates.

deserve their own story.

Author's Note: The search for Mexican War Veterans in the State of Michigan is ongoing. New names will be discovered after the publishing of this book. For updates and the most recent discoveries, please see my virtual cemetery on FindaGrave.com entitled "Mexican War Veterans from Michigan," for more information regarding each, including the obituaries and burial information for these men and women. I also suggest a search of the *Descendents of Mexican War Veterans* website and the *Michigan Sons of Union Veterans* Website.

Next in the Series:

The Second Book in this series will cover all the Regular Army enlistments of Soldiers and Sailors from Michigan.

The Third and final book in this series will cover the entire 1st Michigan Volunteer Infantry Regiment that fought in Mexico.

APPENDIX A:

The Original Company Roster

The Officers:

Major Lewis Cass Jr. survived the war and was discharged, either at St. Louis, Missouri, or at Jefferson Barracks, Missouri, on July 20, 1848.

Captain Andrew T. McReynolds was wounded in action during the fighting around Mexico City and eventually sent home. He was promoted to Brevet Major on August 20, 1847. He was officially discharged July 31, 1848, and received a pension.

First Lieutenant John T. Brown was sent home with the first wagon train to leave Mexico City on November 1, 1847. After some additional time in the hospital at Vera Cruz and New Orleans, he made it home, but no amount of convalescing could help him in the end.

> Death of Lieutenant John T. Brown. -We regret to learn the death of Lieutenant John T. Brown, -He died of a hemorrhage of the lungs at his residence, in Tecumseh, on the 6th inst. Mr. Brown served with distinction in the late war between the United States and Mexico as a Lieutenant under Captain A. T. McReynolds. The army could boast of no braver officer, nor one who did the country more service. He was in several engagements while en route to the city

of Mexico, and at the storming of which, he acquitted himself with lasting honor.

–Adrian Watchtower.[152]

Second Lieutenant John C.D. Williams served for most of the war. After he was wounded in action at Molino Del Rey on September 8, 1847, he remained in Mexico City for a time. Williams was promoted to First Lieutenant on January 8, 1848, replacing Lieutenant Brown. He was eventually sent home to recover from his wounds. On July 31, 1848, he was discharged, most likely at Jefferson Barracks.

Second Lieutenant Francis Henry served with Company K throughout the deployment and remained with the 3rd dragoons after the company was no more, serving in other companies. Presumably he was discharged when the regiment was dissolved at Jefferson Barracks in July of 1848.

The Enlisted Men:

1. **Private Andrew Albro** was a twenty-one-year-old blacksmith from Detroit. He enlisted on March 29, 1847, and mustered into the company on April 22, 1847. He fought with the unit as far as Puebla, where he became ill. He was left in the hospital at Puebla on August 7, 1847. He survived the war and was discharged at an unknown location on July 20, 1848.

2. **Private Charles Allen** was a twenty-three-year-old barber who enlisted in Detroit on March 31, 1847. He mustered in with the company and was present at all of the major battles they fought in. He was transferred to Company C on January 1, 1847. He deserted just a few weeks before he would have been discharged at Jefferson Barracks. His date of desertion was June 28, 1848.

3. **Private Daniel Allen** was a thirty-year-old carpenter from Tecumseh. He enlisted on April 8, 1847, under Lieutenant Brown and mustered in with the company in April but changed his mind during the trip south. He deserted them once the company arrived at New Orleans on May 15, 1847. No further record.

4. **Private Stephen Armstrong** was a twenty-one-year-old blacksmith from Detroit. He enlisted on April 6, 1847, and mustered in

---

[152] Reprinted in the Detroit Free Press on October 10, 1849.

with the company on April 22, 1847. Armstrong was left at the hospital in the city of Puebla on August 8, 1847, after he became sick. On November 5, 1847, he was discharged at Puebla. Armstrong is mentioned in the December 21, 1847, newspaper article as one of those being sent home. He survived the war and had a pension record.

5. **Private Stephen Arnold** was a twenty-two-year-old farmer from Tecumseh. He had been born in Canada and moved to the United States. He enlisted under First Lieutenant Brown on April 7, 1847, and mustered in with the company on April 22, 1847. He was reassigned to Company D on January 1, 1848, until he went home. He was discharged on July 31, 1848, at St. Louis, Missouri. He had an alias he went under that was Stephen H. Johnson.

6. **Private Daniel Axford** was a twenty-nine-year-old Canadian farmer that crossed the river to enlist in Detroit on April 8, 1847. He mustered in with the Company on April 22, 1847, and headed off to Mexico with them. Private Axford was left in the hospital at Puebla on August 7, 1847. Recently discovered evidence suggest that he died there in Puebla on November 9, 1847, having just turned thirty years old a few weeks before. He is buried in the place where all of the dead from that hospital were buried and lies there still.

7. **Private William C. Ayres** was a twenty-one-year-old farmer from Detroit when he enlisted on April 14, 1847. He mustered in with the unit on April 22, 1847, and participated in all the major battles. On October 30, 1847, he was discharged for disability from the hospital in Mexico City and was most likely in the wagon train that headed for Vera Cruz on November 1, 1847. His name was listed in the December 21, 1847, newspaper article as one of those who was discharged and on his way home. He received a pension after the war.

8. **Private Alden Ball** was a twenty-one -year-old farmer from Detroit. He and his twin brother Gilbert enlisted on April 1, 1847, and mustered in on April 22, 1847. Alden served in all the major battles of the war but lost his brother Gilbert to disease at Perote. By October he was also sick and was discharged from the hospital in Mexico City on October 30, 1847. He was most likely with the wagon train of sick and wounded that departed for Vera Cruz on November 1, 1847. Perhaps he was able to visit his brother's grave as they passed through Perote. Alden was mentioned in the December 21, 1847, newspaper article listing those who were coming home to Michigan after discharge.

9. **Private Gilbert Ball** was a twenty-one-year-old farmer from Detroit who had been detached as a teamster in June, ended up in the hospital and died there later, on August 6, 1847. He was buried in a grave outside the along the outside wall of the fort at Perote, Mexico. His twin brother Alden was discharged the following month from a hospital in Mexico City.

10. **Corporal John Barclay** was a twenty-one-year-old carpenter from Canada when he enlisted at Detroit on March 25, 1847. He mustered in with the company on April 22, 1847. He served in the unit until January 4, 1848, when he was transferred to Company C. Barclay was discharged at Jefferson Barracks on July 31, 1848.

11. **Private Ira W. Barlow** was a twenty-year-old miller from Detroit. He enlisted on April 16, 1847, and was mustered into service on April 22, 1847. He went to Mexico and fought alongside them until he became ill. Barlow was left sick in the hospital at Puebla on August 7, 1847. He was discharged on a surgeon's certificate from Puebla on November 5, 1847. Private Barlow received a pension.

12. **Private Franklin Bates** was a twenty-two-year-old farrier from Tecumseh. He enlisted on April 8, 1847, and mustered in on April 22, 1847. He was discharged for disability on October 30, 1847, in Mexico City. Bates was very likely part of the wagon train of sick and wounded that left soon after. He is also very likely the man a newspaper reporter misnamed in the December 21 article as William Batten. No further record.

13. **Private James Beagle** was a twenty-two-year-old painter from Detroit. He enlisted on March 26, 1847, and was mustered into service on April 22, 1847. He fought in all of Company K's battles and was transferred to Company D on January 4, 1848. He was discharged at Jefferson Barracks on July 31, 1848.

14. **Private Joshua Beaman, a.k.a. Bedman** was a twenty-five-year-old farmer from Tecumseh, Michigan. On August 8, 1847, he was left behind at the hospital at Puebla. Sometime after that, he was evacuated back to Jefferson Barracks, Missouri and discharged there on August 1, 1848. Where he was during those 11 months in between is unknown. He received a pension after the war.

15. **Sergeant Edward H. Benedict** was a twenty-six-year-old farmer from Trumbell County, Ohio. He enlisted on April 5, 1847, and mustered in on April 22, 1847. Benedict served throughout the

campaign and was transferred to Company E on January 4, 1848. He was discharged from the army on July 31, 1848, at Jefferson Barracks and received a pension.

16. **Private Clement Beniteau** was a twenty-six-year-old clerk when he enlisted in 1847. Private Beniteau was one of the men left in the hospital at Puebla on August 8, 1847. He was discharged at Jefferson Barracks on August 1, 1848, and received a pension.

17. **Private John Bletter** was a twenty-eight-year-old physician from Baden, Germany when he enlisted in Detroit on March 27, 1847. He was mustered into the company on April 22, 1847. He served during all of the fighting in which the company took part and then was transferred to Company D on January 31, 1847. He was discharged from the army at Jefferson Barracks on July 31, 1847.

18. **Private Charles Burnham**, a twenty-two-year-old "printer" from Detroit. He enlisted on April 16, 1847, and was mustered into the unit on April 22 of that same month. He served with the unit throughout the war and on January 1, 1848, he was transferred to Company D. He was discharged at Jefferson Barracks on 31 July 1848.

19. **Private Thomas Bruette** was a twenty-six-year-old tinsmith from Canada. He enlisted in Detroit on April 22, 1847, the same day the regiment was mustered in. He fought with the company through all of its battles but fell ill. He died in the hospital on September 12, 1847, and is buried near the village of Mixcoac or was interred with the number of dead in the mass grave in Mexico City. Those graves are for the most part unmarked.

20. **Private Charles Burnham** was a twenty-two-year-old printer from Detroit when he enlisted on April 16, 1847. He mustered in on April 22, 1847, and served through all of the battles in Mexico. He was transferred to Company D on January 4, 1848. Burnham was discharged at Jefferson Barracks on July 31, 1848.

21. **Private Elisha Cady** was a twenty-one-year-old from Canada. He enlisted on April 7, 1847, and then vanished back across the river one week later on April 14, 1847. He is listed as a deserter.

22. **Private Alanson Cain** was a twenty-four-year-old distiller from Tecumseh. He enlisted on April 8, 1847, and was mustered into the service on April 22, 1847. Cain was discharged for disability on October 29, 1847, and sent home. He received a pension.

23. **Private William H. Caleb** was a nineteen-year-old farmer from

Tecumseh. He enlisted on April 8, 1847, and was mustered into the service on April 22, 1847. Caleb was left in the hospital at Puebla on July 17, 1847. He was discharged from the Army on July 20, 1848.

24. **Private William Carb**y was a twenty-eight-year-old laborer from Detroit. He enlisted on March 27, 1847, and was mustered in on April 22, 1847. He served with the company in Mexico until January 4, 1848, when he was transferred to Company E.[153]

25. **Private Isaac Carey** was a twenty-six-year-old farmer from Detroit when he enlisted on April 3, 1847. He was mustered in with the other soldiers on April 22, 1847. Carey was left in the hospital at Puebla on August 7, 1847, and he died there on December 12, 1848. He is buried nearby.

26. **Private Samuel Pidesco Cargill** was a twenty-two-year-old clerk from Detroit when he enlisted on April 6, 1847. He mustered in with the company on April 22, 1847, at the Detroit Barracks. He was one of the few to get letters sent home from Mexico, perhaps he was often used as a messenger. He survived all of the battles that the company participated in and was transferred to Company D on January 1, 1848. He became ill soon after and was discharged for disability in Mexico City on February 15, 1848. He was then sent home to Michigan. He received a pension.

27. **Private Thomas Castle** was a twenty-one-year-old waiter from Detroit. He enlisted on March 26, 1847, and was mustered in on April 22, 1847. He fought in all the battles of Company K and then was transferred to Company C on January 4, 1848. He was mustered out at Jefferson Barracks on July 31, 1848.

28. **Private Chandler Ceady or Kailey** was an 18-year-old farmer from Monroe. He enlisted on April 6, 1847, under First Lieutenant Brown at Tecumseh. Ceady, or Kailey mustered in with the unit on April 22, 1847, in Detroit. Private Ceady, or Kailey, died from disease in Mexico City on September 23, 1847, and may be among those interred in the mass grave at Mexico City Possible duplicate in rolls.

29. **Private Charles Clarke** was a twenty-one-year-old blacksmith from Oakland County, Michigan. He enlisted on April 7, 1847, and then deserted with the other deserter, Elisha Cady, on April 14, 1847.

30. **Private Cornelius R. Combs** of Tecumseh was a twenty-three-

---

[153] Pension reference states he died at Sacramento Jun 16, 1884.

year-old farmer when he enlisted under First Lieutenant Brown on April 7, 1847. He mustered in with the company on April 22, 1847, and was killed in action during one of the early engagements of the first convoy on July 6, 1847. Private Combs was buried on the side of the road near the hamlet of Passo Vegas by his friends. He was the first combat casualty of the company.

31. **Private Richard Conklin** was a twenty-one-year-old farmer from Tecumseh. He enlisted on April 6, 1847, and was mustered in on April 22, 1847. He fought in most of Company K's battles and the was transferred to Company C on January 4, 1847. He was discharged on June 31, 1848, at Jefferson Barracks and received a pension.

32. **Private Henry B. Cornwell** was a twenty-two-year-old clerk from Detroit. He enlisted on April 6, 1847, and was mustered in on April 22, 1847. He was left in the hospital at Puebla on August 7, 1847, after falling sick. Cornwell survived and was discharged from the service on July 20, 1848.

33. **Private Daniel Cruice** was a one of the "printers" from Detroit. He enlisted on April 1, 1847, and mustered in with the company on April 22, 1847, at Detroit. He was with the company for all of its major battles. Cruice was too sick to move when the first convoy of sick and wounded was sent to Vera Cruz on November 1, 1847. He died of disease in Mexico City on November 12, 1847, and is buried outside the city. He may be included in the mass grave at the national military cemetery.

34. **Private George C. Dailey** was a twenty-eight-year-old shoemaker from Lower Canada when he enlisted at Tecumseh. He enlisted under First Lieutenant Brown on April 6, 1847, and was mustered in on April 22, 1848. He was left in hospital when the company marched out of Puebla on August 8, 1847. He was discharged from the army on July 20, 1848.

35. **Fourth Sergeant David Davis** was a thirty-four-year-old hosteler from Detroit when he enlisted to serve in Mexico. He enlisted on April 12, 1847, and was mustered in on April 22, 1847. He served with the company until January 4, 1848, when he was transferred to Company G. He was discharged at Jefferson Barracks on August 8, 1848.

36. **Fourth Corporal Erastus M. Denio** was a twenty-two-year-old clerk from Detroit. He enlisted on April 22, 1847, and was mustered into service the same day. He served in Mexico with the company until

he was transferred on January 4, 1848. After that he was a part of Company G until he was discharged on August 8, 1848, at Jefferson Barracks.

37. **Private Charles Diltz** was a twenty-three-year-old coachman from Detroit when he enlisted on April 5, 1847. He was mustered in on April 22, 1847, with the rest of the company at Detroit. He was wounded in action at the Battle of Molino del Rey on September 8, 1847. He was wounded so badly that he was discharged for disability on the same day. He spent time in the hospital at Mexico City until he could be evacuated to the United States. He probably went with the convoy down to Vera Cruz on November 1, 1847. He survived his wounds and received a pension.

38. **Private Patrick Drum** was a twenty-four-year-old farmer from Detroit. He enlisted on April 14, 1847, and was mustered in on April 22, 1847. He served with the company until he was transferred to Company D on January 4, 1848. He was discharged from Jefferson Barracks on July 31, 1848.

39. **Private William Dudley** was a thirty-three-year-old mason from Detroit. He enlisted on March 25, 1847, and was mustered in on April 22, 1847. He served with the company until he was transferred to Company D on January 4, 1848. He was discharged from Jefferson Barracks on July 31, 1848. Private Dudley received a pension.

40. **Private William Eagan** was a twenty-two-year-old farmer from Ireland when he enlisted at Detroit on April 7, 1847. He was mustered in on April 22, 1847. He was transferred to Company C on January 4, 1848. Private Eagan was mustered out at Jefferson Barracks on July 31, 1848.

41. **Private Avery Ellis** was a twenty-five-year-old blacksmith from Tecumseh. He enlisted there under First Lieutenant Brown on April 5, 1847, and mustered in with the company on April 22, 1847, in Detroit. Ellis died of disease in Mexico City on September 25, 1847, and may be among those interred in the mass grave at Mexico City. He never knew what happened to his brother Commodore Ellis.

42. **Private Commodore Ellis** was a twenty-one-year-old farmer from Tecumseh. He enlisted with his older brother Avery on April 5, 1847, under First Lieutenant Brown. Ellis mustered in with the company on April 22, 1847, in Detroit. On the day they landed in Vera Cruz, Ellis was sent to the hospital there, and that is the last he was

heard from. He very likely died there and was buried outside the city walls.

43. **Private William Ellis** was a twenty-three-year-old from Tecumseh and was the last of the three Ellis brothers to still be serving at the end of the war. He enlisted the day after his brothers, April 7, 1847, under First Lieutenant Brown. Ellis served with the company until it was split up on January 4, 1848, and he was transferred to Company C. He was at Jefferson Barracks when he was mustered out on July 31, 1848. His brothers were both buried in Mexico.

44. **Private Michael Fitzgerald** was a twenty-seven-year-old cooper from Ireland when he enlisted at Detroit on April 14, 1847. He was mustered in to the unit on April 22, 1847, and he went to Mexico. Fitzgerald fought in all of the major battles of the 3rd Dragoons and was at Mexico City when he was transferred to Company D on January 4, 1848. He was discharged at Jefferson Barracks on July 31, 1848.

45. **Private Isaac Gibson** was a twenty-one-year-old wagon maker from Detroit. He enlisted on April 5, 1847, and mustered in with the company on April 22, 1847, in Detroit. Gibson fought in all the major battles and on January 1, 1848, was transferred to Company D. He mustered out at Jefferson Barracks on July 31, 1848, and received a pension.

46. **Private William Gibson** was a twenty-seven-year-old carpenter, originally from Ireland, and more recently from Detroit when he enlisted there on April 10, 847. He was mustered in with the rest of the company on April 22, 1847, in Detroit. He was killed in action during one of the many firefights with the rancheros along the convoy route to Mexico City on August 10, 1847. He is most likely buried alongside the road in an unmarked grave.

47. **Corporal Edward Gray** was a twenty-five-year-old farmer from Detroit back in 1847. He enlisted on April 1, 1847, and was mustered in on April 22, 1847. He served with the company until it was disbanded on January 4, 1848, and then was transferred to Company D. This soldier was discharged at Jefferson Barracks on July 31, 1848.

48. **First Sergeant Edward Grey.** This may be the same man as the previous one. His name is slightly different but all other information is identical. He was simply promoted.

49. **Private Nathanial B. Hall** was a twenty-three-year-old farmer

from Detroit. He enlisted on April 1, 1847, and was mustered in on April 22, 1847. He deployed with the company but fell ill on the march. He was left in the hospital at Puebla, Mexico on August 7, 1847.[154] No further record.

50. **Private Herman Harris** was a twenty-one-year-old farmer from Tecumseh. He enlisted on April 5, 1847, under First Lieutenant Brown, and was mustered in on April 22, 1847. He deployed with the company but was left at the hospital at Perote after falling ill on the march on July 3, 1847. No further record.

51. **Private John Heidlauf** was a twenty-nine-year-old farmer originally from Germany, now living in Detroit, He enlisted on April 10, 1847, and was mustered in on April 22, 1847. He served with the company all the way through Mexico until it was dissolved at Mexico City. As of January 4, 1848, he became a member of Company D. He returned to Jefferson Barracks and was discharged on July 31, 1848. His name was sometimes spelled as "Headlauf."

52. **Private Ambrose Hickox** (age unknown) was from Ann Arbor. He enlisted under First Lieutenant Brown on April 7, 1847, at Tecumseh, and was mustered in on April 22, 1847, at Detroit. He deployed with the company but fell ill while on the march. He was left in the hospital at Puebla on August 7, 1847. His name was sometimes spelled "Heacock." No further record.

53. **Private John Hill** was a thirty-five-year-old farmer from Tecumseh. He enlisted under First Lieutenant Brown on April 6, 1847, and was mustered in on April 22, 1847. He fell ill on the march after the Battle at the National Bridge and was left at the hospital at Perote on July 3, 1847. No further record.

54 **Private Ranseller W.**[155] **Hinman**, a 21-year-old farmer from Detroit, He enlisted on April 2, 1847, and was mustered in on April 22, 1847. He fought through all of the battles of Company K. On January 4, 1848, he was transferred to Company C while at Mexico City. He was discharged at Jefferson Barracks on July 31, 1848.

55. **Private Alpheus S. Holloway** was a twenty-nine-year-old

---

[154] His record says June 7, 1847. They left on August 7, 1847.

[155] MEMF has a different spelling of the first name from the Find a Grave entry, the headstone just has his initials. MEMF also states that his middle initial is H., The headstone says W.

carpenter from Tecumseh. He enlisted under First Lieutenant Brown on April 9, 1847, and was mustered in on April 22, 1847. He served with the company in Mexico until he fell ill on the march. He was left in hospital at Puebla on August 7, 1847. No further record.

56. **First Sergeant Julius A. Holmes** was a twenty-four-year-old farmer from Tecumseh. He enlisted on April 5, 1847, under First Lieutenant Brown. He was in charge of mustering in the company at Detroit on April 22, 1847. On October 27, 1847, he was discharged from the service while still at Mexico City and traveled home with some of the sick and wounded.

57. **Private Edwin Howell** was a twenty-nine-year-old carpenter from Detroit. He enlisted on April 9, 1847, and was mustered in on April 22, 1847. He served with the company in Mexico until he fell ill on the march. He was left in hospital at Puebla on August 7, 1847. No further record.

58. **Private Charles P. Hungerford**, a 25-year-old porter from Detroit. He enlisted on April 5, 1847, and was mustered in on April 22, 1847. He served with the unit until it was disbanded at Mexico City. On January 4, 1848, he was transferred to Company D. He was discharged on July 31, 1848, at Jefferson Barracks. Records indicate that he received a pension.

59. **Private Joel M. Jackson** was a twenty-one-year-old farmer from Detroit. He enlisted on April 1, 1847, and was mustered in on April 22, 1847. He served with the company until he became ill. He was left in the hospital at Puebla on August 8, 1847, when the 3rd Dragoons marched on to Mexico City. No further record.

60. **Musician James Johnston** was a thirty-four-year-old carpenter from Detroit. He was originally from Nova Scotia, Canada. He enlisted on April 1, 1847, and was mustered in on April 22, 1847, at Detroit. As a "musician" for the company that would have meant he was the bugler. He became ill while on the march and was taken to the hospital at Puebla shortly after they arrived. Bugler Johnston died in the hospital on July 25, 1847, and the bugle was passed to someone else.

61. **Sergeant John Kennedy**, a thirty-four-year-old farmer from Detroit. He enlisted on March 26, 1847, and was mustered in on April 22, 1847. He served with the unit through all of the major battles but was returned to the ranks as a Private on September 11, 1847, just a few days after the battle at Molino del Rey. The reason is unknown but

if he had been sick, they may have needed to promote someone else to take his place. He was serving as a Private when the company was dissolved, and he was transferred to Company C on January 4, 1848. He was discharged with the others at Jefferson Barracks on July 31, 1848.

62. **Corporal John W. Leake** was a twenty-eight-year-old merchant from Tecumseh when he enlisted under First Lieutenant Brown. He enlisted on April 18, 1847, and mustered in with the company on April 22, 1847. He died from disease on the day the 3rd dragoons left the city of Puebla, on August 7, 1847. He was very likely buried outside the city walls with his comrades.

63. **Private John L. Leavenworth** He enlisted on April 5, 1847, and was mustered in on April 22, 1847. was left in hospital at Perote on July 2, 1847. He may have been evacuated to Vera Cruz. He died "In Hospital" on September 22, 1847. Which hospital he died in is unclear. He may still be buried outside the castle where the hospital was located if he died at Perote.

64. **Private Robert R. Longstreet** was a thirty-year-old teacher from Detroit when he enlisted on April 12, 1847. He mustered in with the company on April 22, 1847. He survived the convoy battles up to Puebla, but after setting out for Mexico City on August 7, 1847, he was "lost on the march" the next day between Puebla and San Martine, on August 8, 1847. He was presumed dead.

65. **Private Alonzo Lurvey** was a twenty-one-year-old cooper from Detroit. He enlisted in Detroit on April 14, 1847, and was mustered in on April 22, 1847. He served with the unit until too sick to travel. He was left in hospital at the city of Puebla on August 7, 1847. There is no further record other than an attempt at a land warrant. He was sometimes called "Alonzo Linvey" in records.

66. **Private Dennis R. Lynch** was a twenty-four-year-old blacksmith from Detroit who served as the blacksmith for the company. He enlisted on April 3, 1847, and was mustered in on April 22, 1847. He served with the unit until it was dissolved and then was transferred to become the blacksmith for Company G at Mexico City on January 4, 1848. He was discharged at Jefferson Barracks on August 8, 1848.

67. **Private Patrick McDivitt** was a twenty-eight-year-old tailor from Detroit. He enlisted on March 26, 1847, and was mustered in on April 22, 1847. He served with the company until it was dissolved and was

transferred to Company C on January 7, 1848. McDivitt was discharged on July 31, 1848, at Jefferson Barracks.

68. **Private Robert G. McIntire** was a twenty-three-year-old mason from Canada. He enlisted on April 6, 1847, and was mustered in on April 22, 1847. He served with the unit until he became ill on the march. He was left on the hospital at Puebla, Mexico, on August 7, 1847. No further record.

69. **Private Edward W. McIntosh** was a twenty-four-year-old farmer from Detroit. He enlisted on April 16, 1847, and was mustered in on April 22, 1847. He went to Mexico with the company but then his history becomes murky. He must have become ill at some point because he was discharged from the service on January 8. 1848 at Vera Cruz, Mexico. This was probably at the general hospital there on the waterfront of the city. No further record.

70. **Private John Marks** was a thirty-three-year-old retired soldier from Detroit. He enlisted on March 20, 1847, and never left Detroit. He became sick at the Detroit Barracks and was put in a hospital nearby. He was not able to be mustered in on April 22, 1847, and died in Detroit the next day. He was buried in the cemetery north of Gratiot that is now the Eastern market. His grave was moved to Elmwood Cemetery north of Jefferson Avenue decades later, and he is buried in one of the many "unknown Soldier" plots in that cemetery.

71. **Sergeant Zeba W. Merritt** was a 24-year-old tinner from Detroit. He enlisted on March 24, 1847, and was mustered in on April 22, 1847. Sgt Merritt served with the company until January 4, 1848, when he was transferred to Company C (or perhaps E.) He was discharged at Jefferson Barracks on July 31, 1848. He received a pension.

72. **Private Stephen Mosher** was a 22-year-old miller from Tecumseh, He enlisted under First Lieutenant Brown on April 8, 1847, and was mustered in on April 22, 1847. He served with the company until it was dissolved at Mexico City on January 4, 1848. After that he was transferred to Company E. He was discharged on July 20, 1848.

73. **Corporal Jesse Myers** was a thirty-year-old lawyer from Detroit, He enlisted on April 10, 1847, and was mustered in on April 22, 1847.He served with the company in Mexico until it was dissolved at Mexico City on January 4, 1848. He was then transferred to Company D. Myers was discharged at Jefferson Barracks on July 31, 1848.

74. **Musician Peter Nelson** was a thirty-one-year-old painter from

Detroit. He enlisted on April 10, 1847, and was mustered in on April 22, 1847. He served in Mexico until he was taken sick at Puebla. He was left in the hospital there on August 7, 1847. At some point he was discharged and sent home. He survived the war and received a pension.

75. **Private David B. Osterhout** was a twenty-one-year-old farmer from Tecumseh. He enlisted on April 1, 1847, and was mustered in on April 22, 1847. He fell ill while encamped at the village of Vergera, outside the city of Vera Cruz. He was left at the hospital there on June 3, 1847. He was discharged with the rest of the company at Jefferson Barracks on July 31, 1848. Private Ousterhout received a pension.

76. **Private Enos Parish**. Enos Parrish was a twenty-one-year-old farmer from Detroit, He enlisted on April 12, 1847, and was mustered in on April 22, 1847. He served with the company throughout the war until it was disbanded in Mexico City on January 4, 1848. He was transferred to Company C. Private Parrish was discharged at Jefferson barracks on July 31, 1848. He received a pension.

77. **Private Joel Parish** was a twenty-seven-year-old lawyer from Detroit. He and his brother Enos were originally from Ontario County, New York. He enlisted on the same day as his brother, April 12, 1847. They were both mustered in on April 22, 1847, at Detroit. Private Parish served with the company until it was dissolved at Mexico City, and then he was transferred to Company C with his brother on January 4, 1848. He did not remain much longer. He was so sick that he was discharged on a surgeon's certificate of disability on January 14, 1848, and sent home.

78. **Private Lorenzo Parsons** was a twenty-four-year-old carpenter from Detroit. He enlisted on April 1, 1847, and was mustered in on April 22, 1847. He served with the unit and then was transferred to Company D when Company K was dissolved on January 4, 1848. He was discharged at Jefferson barracks on July 31, 1848.

79. **Private William H. Patten** was one of the "printers" from Ann Arbor. He enlisted in Detroit on April 5, 1847, and mustered in with the rest of the company on April 22, 1847. He served throughout the campaign and tried to mail something back to the paper when he could. He was discharged for disability in Mexico City on October 30, 1847. He likely was sent home with his fellow printer Welch.

80. **Private William C. Payne** was a twenty-two-year-old farmer

from Cleveland, Ohio. He enlisted at Detroit on April 14, 1847, and mustered in with the rest of the company on April 22, 1847. He served with the company until he was too sick to go on. He was left at the hospital at Puebla on August 7, 1847. No further record.

81. **Corporal Sylvester Peltier**, a nineteen-year-old clerk from Fort Wayne, Indiana. He enlisted at Detroit on March 30, 1847, and mustered in with the rest of the company on April 22, 1847. He served with the company in Mexico until it was disbanded on January 4, 1848, at Mexico City. He was then transferred to Company E. He survived the war and was discharged at Jefferson Barracks on July 31, 1848. He received a pension.

82. **Private Benjamin Raney** was a nineteen-year-old blacksmith from Detroit. He enlisted in Detroit on April 10, 1847, and mustered in with the rest of the company on April 22, 1847. He served with the unit until it was disbanded on January 4, 1848, in Mexico City. He was then transferred to Company D. E. Private Raney survived the war and was discharged at Jefferson Barracks on July 31, 1848. He received a pension. His last name was sometimes spelled Rainey.

83. **Private Justin N. Remington** was a twenty-two-year-old farmer from Detroit, He enlisted at Detroit on March 23, 1847, and mustered in with the rest of the company on April 22, 1847. He served with the company in Mexico until it was disbanded on January 4, 1848, at Mexico City. He was then transferred to Company C. He survived the war and was discharged at Jefferson Barracks on July 31, 1848.

84. **Musician Francis S. Reno** was a twenty-year-old sailor from Detroit, He enlisted at Detroit on March 31, 1847, and mustered in with the rest of the company on April 22, 1847. He served with the company in Mexico until it was disbanded on January 4, 1848, at Mexico City. He was then transferred to Company D. He survived the war and was discharged at Jefferson Barracks on July 31, 1848.

85. **Private Ira Reynolds** was a thirty-three-year-old farmer from Tecumseh. He enlisted under First Lieutenant Brown in Tecumseh on April 9, 1847, and mustered in with the rest of the company on April 22, 1847. He served with the company in Mexico until it was disbanded on January 4, 1848, at Mexico City. He was then transferred to Company C. He survived the war and was discharged at Jefferson Barracks on July 31, 1848.

86. **Private John Riddle** was a twenty-eight-year-old farmer from

Tecumseh. He enlisted under First Lieutenant Brown on April 7, 1847, and mustered in with the rest of the company on April 22, 1847. He served with the company until he was too ill to carry on. He was left behind in hospital at the city of Puebla on August 7, 1847. There was no further record. His name was sometimes written as Rittle.

87. **Private Orville Royce** was a twenty-six-year-old farmer from Tecumseh. He enlisted on April 8, 1847, under First Lieutenant Brown, and mustered in at Detroit on Aprill 22, 1847. Private Royce served in all of the major battles fought by his company until he was too sick to remain in the saddle. He died of disease in Mexico City on October 9, 1847, and may be part of the mass grave at the national military cemetery.

88. **Private Alonzo Seely** was a twenty-two-year-old cooper from Detroit. He enlisted in Detroit on April 2, 1847, and mustered in with the rest of the company on April 22, 1847. He served with the unit until he was too sick to go on. He was left in the hospital at Puebla on August 7, 1847. He survived the war and received a pension.

89. **Private William G. Seymour** of Tecumseh was a twenty-two-year-old farmer when he enlisted under First Lieutenant Brown on April 7, 1847. He mustered in with the company on April 22, 1847. He was wounded during the ambushes of the first convoy and left in the hospital at Puebla on August 8, 1847. He returned to duty shortly thereafter and was the enlisted man likely wounded at the Battle of Molino del Rey, along with Second Lieutenant J. C. D. Williams on September 8, 1847. Seymour mustered out with the company on July 31, 1848, at Jefferson Barracks.

90. **Private James N.A.L (or S.) Simonds was** a twenty-one-year-old farmer from Tecumseh when he enlisted under First Lieutenant Brown on April 5, 1847. He mustered in with the company on April 22, 1847, and proceeded with them to Mexico. He was chosen at Puebla to become one of six soldiers designated as General Scott's personal bodyguard. Simonds was in all of the major battles around Mexico City and was transferred to Company D on January 4, 1848. Simonds mustered out with the unit at Jefferson Barracks on July 31, 1848, and received a pension.

91. **Private John Sly**, the Englishman who enlisted at Detroit, He enlisted at Detroit on April 5, 1847, and mustered in with the rest of the company on April 22, 1847. He served with the unit until too sick

to travel. He died at the hospital in Perote on the 4th of July. Private Sly is still buried outside the castle wall at Perote with the others.

92. **Private Stephen Smith** was a twenty-six-year-old sailor from England. He enlisted in Detroit on April 5, 1847, and mustered in with the rest of the company on April 22, 1847. He served with the unit until he was too sick to travel. He was left in the hospital on August 7, 1848. No further record.

93. **Private Carlisle B. Soper** was a twenty-two-year-old from Clinton County, Ohio. He enlisted under First Lieutenant Brown at Tecumseh on April 8, 1847, and mustered in with the rest of the company on April 22, 1847. He served throughout the war until the company was dissolved on January 4, 1848, in Mexico City. He was then transferred to Company D. He was discharged at Jefferson Barracks on July 31, 1848. Private Soper received a pension.

94. **Private Milton Stoddard** was a twenty-year-old from George County, Ohio. He enlisted at Tecumseh under First Lieutenant Brown on April 8, 1847, and mustered in with the rest of the company on April 22, 1847. He served throughout the war until the company was dissolved on January 4, 1848, in Mexico City. He was then transferred to Company D. He was discharged at Jefferson Barracks on July 31, 1848.

95. **Private Senge Streeter** was a volunteer from Tecumseh. He enlisted under First Lieutenant Brown on April 7, 1847, and mustered in with the rest of the company on April 22, 1847. He served with the unit until he was too sick to travel. He was left at the hospital at Puebla on August 8, 1847. No further record.

96. **Private John Taber** was a twenty-three-year-old laborer from Berkshire County, England. He enlisted in Detroit on either March 27 or April 1, 1847. He mustered in with the rest of the company on April 22, 1847. He served with the unit until it dissolved on January 4, 1848, at Mexico City. He was then transferred to Company D for the remainder of his service. Private Taber was discharged at Jefferson Barracks on July 31, 1848. His name is sometimes spelled Tabor and there is a duplicate in the records.

97. **Private Samuel W. Thorpe** was a twenty-five-year-old farmer from Tecumseh. He enlisted there under First Lieutenant Brown on April 8, 1847, and mustered in with the rest of the company on April 22, 1847. He served in Mexico until he was taken ill. He was left at the

hospital in Puebla on August 7, 1847. No further record.

98. **Private Charles H. Tower** was a twenty-nine-year-old clerk who was originally from Genesee County, New York. He enlisted in Detroit perhaps on impulse to be with his friend Private Cargill. He enlisted the same day the company mustered in on April 22, 1847. He was left behind in the hospital at Puebla when the army left the city on August 7, 1847. After that there is nothing. No further record.

99. **Private William Van Skoik** was a twenty-two-year-old cigar maker from Sackett's Harbor, Michigan. He enlisted in Detroit on April 16, 1847, and mustered in with the rest of the company on April 22, 1847. He served with the company until it was dissolved on January 4, 1848, in Mexico City. Van Skoik was then transferred to Company C. He did not remain there long. He was discharged from the army at the hospital in Mexico City on January 14, 1848, and sent home. No further record.

100. **Private Charles L. Wagoner** was a twenty-four-year-old sailor from Detroit. He enlisted on April 5, 1847, and mustered into the company on April 22, 1847. He was left in the hospital at Vera Cruz on June 3, 1847. He either rejoined the company or was sick for a very long time. He was discharged from Jefferson Barracks with the regiment on July 31, 1848. He was sometimes called Chauncy Wagner.

101. **Private Allen T. (or F.) Welch** was one of the "printers" from Detroit. He was twenty-two when he enlisted on April 16, 1847, in Detroit. He mustered in with the unit on April 22, 1847, and wrote back to the paper when he could. On October 26, 1847, he was discharged for disability from the hospital in Mexico City and traveled back with the November 1, 1847, wagon train to Vera Cruz. He traveled home with Captain McReynolds, who was also being sent home. He received a pension.

102. **Private John D. West**, a twenty-one-year-old clerk from Detroit. He enlisted on April 8, 1847, and was mustered into the company on April 22, 1847. He served in all of the battles in which the company took part. On January 4, 1848, he was transferred to Company D when Company K was dissolved. Private West was mustered out with the regiment at Jefferson Barracks on July 31, 1848.

103. **Private William H. Winters** was a twenty-two-year-old farmer from Detroit. He enlisted in Detroit on April 1, 1847, and mustered in with the rest of the company on April 22, 1847. He went to Mexico but

fell ill almost immediately. He was left in the hospital at Vera Cruz on June 3, 1847, immediately upon their arrival. He must have recovered because he was mustered out with the unit on July 31, 1847, at Jefferson Barracks.

104. **Private Milton A. Wood** was born March 11, 1828. He was a nineteen-year-old farmer from Tecumseh when he enlisted on April 7, 1847, under First Lieutenant Brown. He mustered in with the company on April 22, 1847, in Detroit. After the Battle at the National Bridge, he became ill, and died of disease on July 3, 1847, in the hospital at Perote. He is buried in the dry moat on the south side of the wall with the other unmarked graves.

Brothers in Arms:

Private Albert G. Barnes. He was twenty-one-years-old when he enlisted in New York and served in Company E of the 3rd Dragoons under Captain Duff and then First Lieutenant Diwer. He was in all of the same battles as Company K except the first convoy. Barnes came to Michigan after the war in 1848 because his family had moved there, and he settled in Algensee Township, Branch County, Michigan. He spent the rest of his life there. A Branch County newspaper reported that, *"Albert Barnes, a veteran of the Mexican war, died at his home in Algensee, Monday night."*[156]

Private William Atkinson. Since this is a book about Michigan, he is worth mentioning. Atkinson was born in 1821 and was twenty-five years old when he joined the army. He was not in the 3rd Dragoons, but he was a Corporal in the 2nd Dragoons, and he said that he was from Detroit, Michigan when he enlisted for five years of service on July 26, 1846, in Baltimore, Maryland under Captain Croghan Ker. It appears he signed up as a replacement for those who were lost in the early battles of the war. He saw most of the same service that the men of Company K did. He served in Company K of the 2nd Dragoons and was then transferred to Company F when the companies were consolidated. He was discharged at the City of Mexico, the date is illegible, suggesting the reason was illness and disability or wounds. His widow received a pension after his death.

---

[156] The Courier and Republican, Coldwater, June 3, 1898.

Captain Daniel Henry Rucker. In 1837 he received a commission as a 2nd Lieutenant in the 1st Dragoons Regiment when he was twenty-five-years-old. He was from Grosse Isle, Michigan, and received his commission in Detroit. His uncle was Alexander Macomb, commanding general of the United States Army after the War of 1812.

He also served in the Mexican War. He was in command of Company E of the 1st Dragoons that was attached to General Taylor's army in the northern theater of war. He distinguished himself at the Battle of Buena Vista and earned a Brevet Major's position. After the war he transferred to the Quartermaster's Corps, where he served during the Civil War, and beyond. At the end of his career in 1882, he had been promoted to Brigadier General and was the Quartermaster General of the United States Army.

APPENDIX B:

The Recruits.

Private John Blackmer, a thirty-four-year-old farmer who enlisted in Detroit on April 24, 1848. He was discharged on July 18, 1848, at Jefferson Barracks.

Private Barton Bowdine, an eighteen-year-old Canadian farmer who enlisted in Detroit on April 21, 1848. He was discharged at Jefferson Barracks on July 16, 1848, and received a pension.

Private Porter Briggs was a twenty-one-year-old blacksmith who enlisted in Detroit on June 3, 1848. He was discharged at Jefferson Barracks on July 16, 1848.

Private James B. Darby was a twenty-one-year-old farmer on May 9, 1848, when he enlisted as one of McReynold's replacements. He died at Newport Kentucky and therefore never even made it to Jefferson Barracks. He was most likely buried in Newport Kentucky.

Private Joseph B. Davidson was a twenty-four-year-old farmer from Detroit who enlisted on May 12, 1848, in Detroit. He was discharged at Jefferson Barracks on July 16, 1848, and received a pension.

Private Michael Duke was a twenty-five-year-old laborer from Ireland. He enlisted in Detroit on May 15, 1848, and then deserted sometime around July as the others were being discharged.

Private Michael Fitzsimmons was a Great Lakes sailor who enlisted at Detroit on May 10, 1848. He was discharged on July 16, 1848, at Jefferson Barracks.

Private Edward R. Foy was recruited at Detroit on May 12, 1848. On July 16, 1848, he was discharged on July 16, 1848, at Jefferson Barracks.

Private Robert Gordon enlisted in Detroit on April 21, 1848. He was an eighteen-year-old farmer from the area. On July 16, 1848, he was released from service at Jefferson Barracks.

Private John Haight, a twenty-one -year-old carpenter from Detroit, enlisted on May 2, 1848. He deserted on May 24th, 1848.

Private Konrad Hajker was a twenty-one -year-old immigrant shoemaker from Germany when he enlisted in Detroit on April 20, 1848. He was discharged from Jefferson Barracks on July 16, 1848.

Private Samuel Hancock was a twenty-one -year-old farmer from the Detroit area who enlisted on May 20, 1848. He was discharged on July 16, 1848, at Jefferson Barracks.

Private John Hickey was a twenty-two-year-old Irish Laborer who enlisted in Detroit on May 22, 1848. He was discharged on July 16, 1848, at Jefferson Barracks.

Private Edward Hughes was a twenty-one-year-old Irish laborer who enlisted in Detroit on April 19, 1848. He was discharged on July 16, 1848, at Jefferson Barracks.

Private John B. Hyatt was a thirty-three-year-old molder from Nova Scotia who had recently come to Detroit. He enlisted in Detroit on May 16, 1848. He was discharged on July 16, 1848, at Jefferson Barracks.

Private Samuel Loomis was a twenty-one-year-old farmer that was recruited in Detroit on June 5, 1848. 35 days later he was discharged on July 10, 1848, at East Pascagoula, Mississippi.

Private James McClellan was a thirty-two-year-old farmer who enlisted in Detroit on May 31, 1848. He was discharged on July 16, 1848, at Jefferson Barracks.

Private James McFarlan was a twenty-four-year-old cooper who enlisted in Detroit on May 20, 1848. He was discharged on July 16, 1848, at Jefferson Barracks.

Private John McKnugent was a teacher from Canada who enlisted in Detroit on May 9, 1848. He was discharged on July 16, 1848, at Jefferson Barracks.

Private John Nugent was also a teacher from Canada who enlisted in Detroit on May 9, 1848. He was discharged on July 16, 1848, at Jefferson Barracks. This may be a typo of the above soldier.

Private James Preston was a thirty-four-year-old tailor from Ireland. He enlisted on May 16, 1848. He was discharged on July 31, 1848, at Jefferson Barracks.

Private Benjamin Rose was a thirty-four-year-old sailor who enlisted in Detroit on April 26, 1848. He was discharged on July 16, 1848, at Jefferson Barracks.

Private William Rose was a thirty-four-year-old carpenter who enlisted in Detroit on April 27, 1848. He was discharged on July 16, 1848, at Jefferson Barracks.

Private Francis Salute was a twenty-five-year-old laborer from Germany who enlisted on May 29, 1848. He was discharged on July 16, 1848, at Jefferson Barracks.

Private Henry R. Wilson was a thirty-four-year-old cooper who enlisted at Detroit on May 30, 1848. He was discharged on July 16, 1848, at Jefferson Barracks.

Private George Wright was a twenty-four-year-old sailor who enlisted in Detroit on May 2, 1848. He was discharged on July 16, 1848, at East Pascagoula, Mississippi.

Private Jackson Wright was a twenty-one-year-old sailor who enlisted in Detroit on May 2, 1848. He was discharged on July 16, 1848, at East Pascagoula, Mississippi.

BIBLIOGRAPHY

Books:

Le Roy Barnett & Roger Rosentreter, *Michigan's Early Military Forces: A Roster and History of Troops Activated Prior to the American Civil War*. Wayne State University Press, Detroit, 2003.

Bauer, Jack K. *The Mexican War 1846-1848*. Lincoln, University of Nebraska Press, Lincoln and London, 1974.

Bracket, Albert G., *General Lane's Brigade in Southern Mexico*. H. W. Derby & Co., Cincinnati, Ohio, 1854.

Bonner, Richard Illenden, editor. *Memoirs of Lenawee County, Michigan. Volume I*. Madison: Western Historical Association, 1909.

Christenson, Thomas and Carol, *The U. S. Mexican War:* Companion to the Public Television Series. Bay Books, San Francisco, 1998.

Crawford, Mark, *Encyclopedia of the Mexican War*. ABC-CLIO. Denver, Colorado, 1999.

Farmer, Silas. *History of Detroit and Wayne County and Early Michigan. Volume I*, 3rd edition. Revised, Detroit, Silas Farmer & Company, 1890.

Field, Ron. *Mexican-American War 1846-1848:* Brassey's History of Uniforms. Brassey's (U.K.) Limited: 1997.

History of Shiawassee and Clinton Counties, Michigan. Philadelphia: D. W. Ensign and Company. 1880.

History of Genesee County, Michigan. Philadelphia, Everts & Abbott, 1879.

Hughes, N. C., and Timothy Johnson., ed. *A Fighter from Way Back: The Mexican War Diary of Lieutenant Daniel Harvey Hill, 4th Artillery, U.S.A.,* Kent State University Press, Kent, Ohio: Kent State University Press, 2002.

Johannson, Robert W., *To the Halls of the Montezumas: The Mexican War in the American Imagination*. Oxford University Press, New York, 1985.

Johnson, Timothy D. *A Gallant Little Army: The Mexico City Campaign.* University Press of Kansas, 2007.

Lambert, Joseph I., "One Hundred Years with the Second Cavalry." Capper Printing Company, Inc. Fort Riley, Kansas. 1939.

Nevin, David, *The Mexican War*: From the "Old West" Series, Time-Life

    Books, Alexandria, Virginia, 1978.

Orr, William J. Orr & Robert Ryal Miller, Editors. *Frederick Zeh: An*

    Immigrant in the Mexican War, Translated by College

    Station: Texas A&M University Press, 1995.

Ripley, R. S., *The War with Mexico*. 2nd Volume, New York: Harper &

    Brothers, 1849.

Scott, Robert Garth, ed. *Forgotten Valor: The Memoirs, Journals, & Civil*

    War Letters of Orlando B. Willcox. Kent State University

    Press, Kent, Ohio & London, 1999.

Seeley, Thaddeus D. *The History of Oakland County, Michigan*. Volume

    I., Chicago: The Lewis Publishing Company, 1912.

Smith, Justin Harvey, The War with Mexico, Volume II, 1846-1848.

    New York, The Macmillan Company, 1919.

Steffen, Randy, *The Horse Soldier 1776-1943, Volume I; The Revolution,*

    *The War of 1812, the Early Frontier, 1776-1850.*Universiy of

    Oklahoma Press, Norman, 1977.

Weems, John Edward, *To Conquer a Peace: The War Between the United*

States and Mexico. Doubleday and Company, Garden City,

New York, 1974.

Wing, Talcott E., editor. *History of Monroe County, Michigan.* New York:

Munsell & Company, 1890.

Articles:

Hamersley, L. R. Ed., "General Lane, "Our Cavalry in Mexico." *The*

United Service Magazine. (June 1896): 487.

McReynolds, Andrew. "Presidential Address to the "Veterans of the

War with Mexico." *Pioneer and Historical Society,*

Collections. Lansing: W. S. George & Co., (1884,) VI, 20.

Moseley, W. G. "Reminiscences of a Mexican Campaign." *The Vedette.*

(January-June 1880): 1-19.

Gibson, Isaac. "Phil Kearny's Charge: An Interesting Letter from

Michigan." *The Vedette.* Volume 1. (April 15, 1880): 12-13.

Ray, O. L. "Reminiscences of Scott's Campaign." *The Vedette.* Volume 1.

(April 15, 1880): 12-13.

Newspapers:

Editor, *Coopersville Observer*, (Allendale, MI.) April 25, 1902.

Editor, *Courier and Republican*, (Coldwater, MI.) June 3, 1898.

Editor, *Detroit Daily Advertiser*, (Detroit, MI.) 1846-1848.

Editor, *Detroit Democratic Free Press*, (Detroit, MI.) April 12, 1847.

Editor, "The Michigan Dragoons." *Detroit Democratic Free Press*, (Detroit MI.) April 24, 1847.

Editor, "Presentation of a Sword to Captain McReynolds." *Detroit Democratic Free Press*, (Detroit, MI.) Apr. 30, 1847.

Welch, Allen T., *Detroit Democratic Free Press*, (Cincinnati, OH.) May 8, 1847.

Cargill, Samuel P., *Detroit Democratic Free Press*, (Vera Cruz, MX.) July 8, 1847.

Editor, *Detroit Democratic Free Press*, (Detroit MI.) Dec. 9, 1847

Editor, "Captain McReynold's Charge." *Detroit Democratic Free Press*, (Detroit, MI.) Dec. 15, 1847.

O. B. Holman, "By Telegraph from Ypsilanti" *Detroit Democratic Free Press*, (Ypsilanti, MI.) Dec. 10, 1847.

Editor, *Detroit Evening News*, (Detroit, MI.) Sept. 25, 1895.

Editor, "He was a Man Among Men: Death of Gen. A. T. McReynolds, Who Helped Make U. S. History." *Detroit Free Press*, (Detroit, MI.)

Nov. 26, 1898.

"The Third Dragoons, A Popular Company." *Jackson Patriot*, (Jackson,

MI.) Aug. 31, 1847.

Editor. *Jackson Patriot*, (Jackson, MI.) Dec. 7, 1847.

Editor, *Kalamazoo Gazette*, (Kalamazoo, MI.) April 21, 1848.

Editor, "Arrival of Troops." *Hillsdale Whig Standard*, (Hillsdale, MI.)

May 18, 1847; December 21, 1847.

Cargill, Samuel P. "Michigan Boys in Mexico." *Hillsdale Whig Standard*,

(Hillsdale, MI.) July 13, 1847.

Editor, "Letter from Mexico-[regarding] Lieutenant J. T. Brown."

Hillsdale Whig Standard, (Hillsdale, MI.) Dec. 28, 1847.

Editor, *Niles National Register*, November 23, 1847.

Editor, *Oakland Gazette*, (Pontiac, MI.) July 31, 1847.

Online Sources:

Ancestry.com Website. https://www.ancestry.com

The Library of Congress. Home | Library of Congress

Michigan Sons of Union Veterans of the Civil War. Department of

Michigan - Sons of Union Veterans of the Civil War

National Archives (NARA) online. **National Archives | Home**

Thesis and Compilations:

Smith, Kenneth A. "Michigan's Military Participation in the Mexican War." M. A. Thesis, Wayne State University, 1951.

McReynolds, Andrew, "Presidential Address to the Veterans of the War with Mexico." Michigan Pioneer and Historical Society, Collections. Lansing: W. S> George & Company, 1884, Volume VI.

Richard W. Welch, Compiler, "Michigan in the Mexican War." From Record Group 59-14, Records of the Michigan Military Establishment.

Records pertaining to the Mexican War 1846-1848. In the Michigan Historical Commission Archives, Lansing, Michigan. December 15, 1967. U.S. Michigan, M2W.

Specific Government Documents:

The National Archives Title: Registers of Enlistments in the United States Army 1798-1914. (M-233.) [Fold-3 Electronic Source Version]

The National Archives Title: US, Letters Received by the Office of the

Adjutant General Main Series 1822-1860 (Fold-3 Electronic

Source Version]